The Golem at Large: what you should know about technology

In the very successful and widely discussed first volume in the Golem series, *The Golem: What You Should Know About Science*, Harry Collins and Trevor Pinch likened science to the golem, a creature from Jewish mythology, a powerful creature which, while not evil, reveals a flailing and clumsy vigour. Through a series of fascinating case studies of famous and not-so-famous scientific episodes, ranging from relativity and cold fusion to memory in worms and the sex lives of lizards, the authors showed that science too is neither all good nor all bad. They debunked the traditional view that science is the straightforward result of competent theorization, observation and experimentation; they showed it is a set of fallible skills.

In this second volume, the authors now consider the golem of technology. Using the same successful format, in a series of case studies they demonstrate that the imperfections in technology are related to the uncertainties in science described in the first volume. The case studies cover the role of the Patriot anti-missile missile in the Gulf War, the *Challenger* Space Shuttle explosion, tests of nuclear fuel flasks and of anti-misting kerosene as a fuel for airplanes, economic modelling, the question of the origins of oil, impact of the Chernobyl nuclear disaster and the contribution of lay expertise to the analysis of treatments for AIDS.

The Golem series tries to build an island between the two cultures of science on the one hand and the humanities and social sciences on the other. It is an attempt to explore science as a product of social life and to demonstrate that the study of science is the proper concern of more than just scientists. This fascinating work will be of interest to readers from a wide range of science and non-science backgrounds. It could be used to give science students a humanistic interpretation of their work, but also to introduce non-scientists to the analysis of science. Anyone who found the original Golem volume of interest will want to read this latest volume.

Harry Collins is Distinguished Research Professor in Sociology and Director of the Centre for the Study of Knowledge, Expertise and Science (KES) at Cardiff University. His books include *Changing Order: Replication and Induction in Scientific Practice* (1985, Second edition published by Chicago University Press in 1992), *Artificial Experts: Social Knowledge and Intelligent Machines* (1990), published by MIT Press, *The Shape of Actions: What Humans and Machines Can Do* (MIT Press with Martin Kusch) and, with T. Pinch, *The Golem: What You Should Know About Science* (Cambridge University Press, 1993), which won the 1995 Robert Merton Prize of the American Sociological Association. He is the 1997 recipient of the J. D. Bernal Award of the Society for Social Studies of Science. He is part way through a new study of the history and sociology of gravitational wave physics.

Trevor Pinch was born in Ireland and educated in the UK. He has degrees in physics and in sociology. He taught sociology at the Universty of York before moving to the USA. He is currently Professor in the Department of Science and Technology Studies, Cornell University. He has held visiting positions at the University of California San Diego, The Danish Technical University, the Science Centre Berlin and the Max Planck Institute for the History of Science, Berlin. His main research area is in the sociology of science and technology where he is the author of several books including *Confronting Nature: The Sociology of Solar-Neutrino Detection*, Kluwer, 1986; *Health and Efficiency: A Sociology of Health Economics* (with Malcolm Ashmore and Michael Mulkay), Open University Press, 1989 and co-editor of *The Social Construction of Technological Systems* (with Wiebe Bijker and Tom Hughes), MIT Press, 1987 and *The Handbook of Science and Technology Studies* (with Sheila Jasanoff, Gerald Markle and James Petersen) Sage, 1995. His book *The Golem: What Everyone Should Know About Science* (with Harry Collins) CUP, 1993 won the Merton Prize of the American Sociological Association. He also researches the sociology of markets, where he is the author (with Colin Clark) of *The Hard Sell*, HarperCollins 1996. His current research is on the history of the electronic music synthesizer.

The Golem at Large

what you should know about technology

Harry Collins
Trevor Pinch

CAMBRIDGE
UNIVERSITY PRESS

PUBLISHED BY THE PRESS SYNDICATE OF THE UNIVERSITY OF CAMBRIDGE
The Pitt Building, Trumpington Street, Cambridge CB2 1RP, United Kingdom

CAMBRIDGE UNIVERSITY PRESS
The Edinburgh Building, Cambridge CB2 2RU, UK
40 West 20th Street, New York, NY 10011-4211, USA
477 Williamstown Road, Port Melbourne, VIC 3207, Australia
Ruiz de Alarcón 13, 28014 Madrid, Spain
Dock House, The Waterfront, Cape Town 8001, South Africa

www.cambridge.org

First published 1998
Canto edition 2002

Printed in the United Kingdom at the University Press, Cambridge

Typeset in Sabon

A catalogue record for this book is available from the British Library

ISBN 0 521 55141 2 (hardback)
ISBN 0 521 01270 8 (paperback)

Cover illustration: *Julia Fractal*, Mehau Kulyk/Science Photo Library. Computer-generated fractal derived from the Julia Set. Fractals are patterns that are formed by repeating some simple process on an ever decreasing scale. The Julia Set is a class of shapes plotted from complex number coordinates. It was invented and studied during World War I by the French mathematicians Gaston Julia and Pierre Fatou. Fractals are used to model natural forms and to study chaos theory.

For
SADIE COLLINS
and to the memory of
OWAIN PINCH

Contents

Preface and acknowledgements

With the exception of most of Chapter 1 and the whole of Chapter 3, the substantive parts of this book are largely expositions of others' work; in this we follow the pattern of the first volume in the Golem series. The full bibliographic references to the works discussed both in this Preface and the other chapters, as well as additional reading, will be found in the Bibliography at the end of the volume.

As for the substantive chapters, Chapter 1 is Collins's redescription of the argument over the success of the Patriot missile. It is heavily based on the record of a Congressional hearing that took place in April 1992, and on two papers written by principal disputants, Theodore Postol and Robert Stein; it also draws on wider reading. Though this chapter is not a direct exposition of anyone else's argument, and though it uses a new analytic framework turning on different definitions of success, it must be made clear that the account was made possible only because of Postol's prior work. Also, Postol was extremely generous in supplying Collins with much of the relevant material and drawing his attention to more. Collins has tried to make sure that the account is not unduly influenced by Postol's views and that the material on which it draws represents the field in a fair way. It will be noted that the chapter does not repeat Postol's expressed position – that no Scud warheads, or almost no Scud warheads, were destroyed by Patriot missiles – but stresses the difficulty of reaching any firm conclusion while keeping open the strong possibility that Postol is right.

Chapter 2 is based on parts of Diane Vaughan's book, *The Challenger Launch Decision: Risky Technology, Culture and Deviance at NASA*. In the context of earlier work, Pinch had previously read McConnell's book, *Challenger: 'A Major Malfunction', Report of the Presidential Commission on the Space Shuttle Challenger Accident* and Gieryn and Figert's article, 'Ingredients for a Theory of

Science in Society'. An earlier article by Pinch on this topic can be found in the Bibliography, but Vaughan's meticulous research provides a detailed historical ethnography of the events leading up to the *Challenger* launch decision and leads to a new interpretation of them.

Chapter 3 is an exposition and simplification of one of Collins's own papers, 'Public Experiments and Displays of Virtuosity: The Core-Set Revisited' which is referred to in the Bibliography.

Chapter 4 is based on Simon Cole's article, 'Which Came First, the Fossil or the Fuel?' This article was originally prepared as a term paper for a course taught by Pinch at Cornell University. Pinch also has access to an unpublished article by Cole on the Gold affair. We thank Bill Travers and Bill Travers Jr. for reading an earlier draft of the chapter.

The substantive part of Chapter 5, 'Tidings of Comfort and Joy?', is based on Collins's reading of Robert Evans's paper 'Soothsaying or Science: Falsification, Uncertainty and Social Change in Macroeconomic Modelling'. Each quotation is taken from Evans's discussions with economists. Collins was also able to take advantage of a great deal of discussion with Evans who was a doctoral student at the University of Bath between 1992 and 1995, working under his supervision. The 'Soothsaying or Science' paper is based on this PhD, which is referred to in the Bibliography.

It should not be thought that all economists are unaware of the problems and ironies of their own discipline as the books by Ormerod and Wallis reveal. The Reports of the British Government's 'Panel of Independent Forecasters' (The Seven Wise Men) for February 1993 and February 1994 were also an important source of information for the chapter. Collins also used ideas from his own 'The Meaning of Replication and the Science of Economics'.

It should be noted that the 'Discussion' and 'Postscript' sections of Chapter 5, which include some speculations, are essentially the responsibility of the authors of this volume though most of the sceptical quotations are taken from economists.

For Chapter 6, Pinch used several articles by Brian Wynne on the effect of Chernobyl radiation on Cumbrian sheepfarmers. These articles are referred to in the Bibliography.

Chapter 7 is based upon Pinch's reading of Steven Epstein's book

Impure Science: AIDS, Activism and the Politics of Knowledge and an article, 'The Construction of Lay Expertise'.

In the case of all chapters, both Pinch and Collins read and re-worked the drafts and take joint responsibility for the results.

We are extremely grateful for the generosity and the efforts of the authors of the works on which we have based our accounts. In every case they were unstinting with their time and effort and, by reading our versions of their work, helped us make sure that we did not stray too far from their intentions. Wiebe Bijker and Knut Sorensen read and advised on the Introduction and the Conclusion. We thank Park Doing for suggesting the words 'at Large' in our title. That said, the final responsibility for mistakes in exposition, infelicities of style, and errors of judgement or analysis, remains our own.

Introduction: the technological golem

'Science seems to be either all good or all bad. For some, science is a crusading knight beset by simple-minded mystics while more sinister figures wait to found a new fascism on the victory of ignorance. For others it is science which is the enemy; our gentle planet, our slowly and painfully nurtured sense of right and wrong, our feel for the poetic and the beautiful, are assailed by a technological bureaucracy – the antithesis of culture – controlled by capitalists with no concern but profit. For some, science gives us agricultural self-sufficiency, cures for the crippled, a global network of friends and acquaintances; for others it gives us weapons of war, a school teacher's fiery death as the space shuttle falls from grace, and the silent, deceiving, bone-poisoning, Chernobyl.

Both of these ideas of science are wrong and dangerous. The personality of science is neither that of a chivalrous knight nor pitiless juggernaut. What, then, is science? Science is a golem.

A golem is a creature of Jewish mythology. It is a humanoid made by man from clay and water, with incantations and spells. It is powerful. It grows a little more powerful every day. It will follow orders, do your work, and protect you from the ever threatening enemy. But it is clumsy and dangerous. Without control a golem may destroy its masters with its flailing vigour; it is a lumbering fool who knows neither his own strength nor the extent of his clumsiness and ignorance.

A golem, in the way we intend it, is not an evil creature but it is a little daft. Golem Science is not to be blamed for its mistakes; they are our mistakes. A golem cannot be blamed if it is doing its best. But we must not expect too much. A golem, powerful though it is, is the creature of our art and our craft.'

This extract from the first volume of the series, *The Golem: What You Should Know About Science*, explains why we chose a golem as our motif.[1] Now we turn from science to its applications; things are not so different.

Like its predecessor, this book contains seven stories. These are the themes: it is very hard to say whether Patriot missiles succeeded in shooting down Scuds during the Gulf War; the blame for the explosion of the *Challenger* Space Shuttle is much less easy to assign than it is usually taken to be; conclusions about safety drawn from deliberate crash-tests of a train and an airplane were less clear than they appeared; the origins of oil are more controversial than we thought and surprisingly difficult to pin down; economic models built by the British government advisors have such big uncertainties that they are useless for forecasting; the consequences of the radioactive fallout from Chernobyl were misunderstood by the official experts; and research on a cure for AIDS needs the expertise of patients as well as doctors and researchers.

The stories, we hope, are interesting in themselves, but their full significance must be understood in terms of the escapades of golem science. The problems of technology – we use the term loosely to mean 'applied science' – are the problems of science in another form.

As with *The Golem*, *The Golem at Large* has a very simple structure. The substance is in the stories but we also draw out the consequences in a short conclusion. We do not expect the conclusion to carry conviction without the substance. While the basis of the argument is drawn from the history and sociology of scientific knowledge, we have introduced only a few technical principles from that field. As in the first volume, we have made considerable use of the notion of the 'experimenter's regress'. This shows that it is hard for a test to have an unambiguous outcome because one can never be sure whether the test has been properly conducted until one knows what the correct outcome ought to be.

Another idea introduced in this volume is the notion that in science and technology, as in love, 'distance lends enchantment'. That is to say, scientific and technological debates seem to be much more simple and straightforward when viewed from a distance. When we find ourselves separated from our loved-ones we remember only why we love them; the faults are forgotten. In the same way, science and technology, when understood through others' inevitably simplified accounts, look artless. Closer to the centre of a heated debate, the less pre-determined and more artful do science and technology appear. The irony is that, quite contrary to what common-sense might lead one to expect, it is often that the greater one's direct experience of a case, the less sure one is about what is right.

A third idea, made much use of in the chapter on the Patriot missile, is the notion of 'evidential context'. It shows that the meaning of the same experimental finding, or test outcome, can seem positive or negative depending upon the problem that the finding is taken to address.

In each story we have tried to display technical details in as simple a way as possible. Even if the overall conclusion of the book is resisted we hope to explain some technology to those who might not otherwise encounter it. Mostly we tell of technological heroism but we present it as a human endeavour rather than a superhuman feat. Technology should not be terrifying or mysterious, it should be as familiar as the inside of a kitchen or a garden shed; as we will see, the decisions made by those at the frontiers are, in essence, little different from those made by a cook or a gardener.

Technology is demonstrated and used in conditions which are under less control than is found in scientific laboratories. As we will see in this book, those faced with the uncertainties of technology are inclined to look toward the controlled environment of science as a golden solution. But science cannot rescue technology from its doubts. The complexities of technology are the same as those that prevent science itself from delivering absolutes; an experimental apparatus is a piece of technology and, looked at closely, the conditions seem as wild inside the lab as outside. Both science and technology are creatures of our art and our craft, and both are as perfectible or imperfectible as our skill allows them to be. As we showed in *The Golem*, science at the research frontier is a matter of skill, with all the uncertainty that that implies; technology at the frontier is the same.

While technologists dream of the perfection of science, the reliability of everyday technology is often taken to prove the enduring infallibility of science. Rockets go to the moon, airplanes fly at 30,000 feet and the very word-processor on which this book has been typed seems to be a tribute to the irrevocability of the theories used in its design. There is something suspicious about this argument too. Why do scientists who, as they see it, work in maximally controlled conditions, turn away from the laboratory to verify their position in the world and the truth-like character of their enterprise? And, if technology is the staff of science, why is it that when technology fails – as in the Chernobyl melt-down and the Space Shuttle explosion – science does not fail? Furthermore, how is it that superb technologies like the barrel and the waggon wheel seemed to lead a life independent of science? The argument from the reliability of technology has a 'no-lose' clause: Chernobyl and the Space Shuttle can verify science if they work but cannot damage it if they do not. The clause is enforceable because failures of technology are presented as failures of human organization, not science.

When science seems less than sure, technology is cited in its defence, and when technology seems less than sure, science is summoned to the rescue; the responsibility is passed backward and forward like the proverbial hot potato. And if the potato is dropped it is always people who are said to drop it. Our picture of the relationship is far more straightforward. Both science and technology are skilful activities and it cannot be guaranteed that a skill will always be executed with precision. Technology is not the guarantor of science any more than science is the guarantor of technology.

This is not to say that we advise the readers of this book to worry more when they board an airplane. Just as science may enter a realm where no one questions or disputes its findings, so technology becomes more reliable as our experience grows and our abilities develop. The workings of gravitational wave detectors and solar neutrino counters are still matters of dispute; the workings of voltmeters and small, earth-bound telescopes are not. The workings of Space Shuttles and AIDS cures are still matters of dispute; the working of barrels, waggon wheels, personal computers, and motor cars are not.

It would, of course, be foolish to suggest that technology and science are identical. Typically, technologies are more directly linked to the worlds of political and military power and business influence than are sciences. Thus, as we show, the outcome of the argument over the success of the Patriot anti-missile missile links directly to the economic fortunes of the firms that make it and to the military postures of governments. Relationships of a similar kind can be seen in respect of every case discussed in this book, whereas in the first volume of the series the distance between the science and the national and business environments was far greater. But these are difference of degree; the influences of military, political, economic, and other social forces may give obvious vigour and longevity only to certain technological debates, but the potential of these forces is found in the structure of scientific and technological knowledge. Since all human activity takes place within society, all science and technology has society at its centre.

The book can be read in any sequence but the end of the volume stresses the contributions of 'lay experts' more heavily. The idea of golem science and golem technology does not imply that one person's view is as good as another's when it comes to scientific and technical matters; on the contrary, the Golem series turns on the idea of expertise. But where is the border of expertise? What is clear from the latter cases is that the territory of expertise does not always coincide with territory of formal scientific education and certification.

The examples of Cumbrian sheepfarmers and AIDS patients presented in this book are cases where people normally referred to as laypersons made a vital contribution to technical decisions, but these are only laypersons in the sense of their not possessing certificates. In fact the farmers were already experts in the habits of sheep and the flow of water on the Cumbrian fells, and the AIDS patients already knew most of what could be known about the habits of AIDS sufferers, their needs and their rights. Furthermore, as time went on, many of the AIDS patients trained themselves to become fluent in the language and concepts of medical research. Bringing such persons into the technological decision-making process should not be seen simply as a democratic necessity; rather it is good sense in terms of using available expertise even when it is found in unexpected places.

In our cases, undemocratic reflexes may have delayed the recognition of this 'lay-expertise', but the solution was not just more democracy. After all, what expertise does a member of the public, as a member of the public, bring to technological decision making? Expertise is too precious for the problem of its recognition to be passed wholly into the sphere of politics. Lay political activism may sometimes be necessary to jerk people out of their comfortable assumptions about the location of expertise, but success in this sphere makes it all too easy to jump to the conclusion that expertise can be replaced with heartfelt concern; this is wrong.

Authoritarian reflexes come with the tendency to see science and technology as mysterious – the preserve of a priest-like caste with special access to knowledge well beyond the grasp of ordinary reasoning. It is only through understanding science and technology as golem-like – as failure-prone reachings out of expertise into new areas of application – that we will come to understand how to handle science and technology in a democratic society and resist the temptation to lurch from technocracy to populism.

Finally, let us be clear that we look only at cases where conclusions are the subject of dispute. Such events are a statistically unrepresentative sample of science and technology because most science and technology is undisputed. But disputes are representative and illustrative of the roots of knowledge; they show us knowledge in the making. As we journey forward into the technological future we see the technological past as through mirrors barring our route. These backward-looking mirrors provide a distorted perspective in which everything is seen as settled almost before it was thought about; *The Golem at Large* is meant to lead us from the hall of mirrors.

NOTE

[1] In fact the first volume was called *The Golem: What Everyone Should Know About Science*. The title has been slightly changed for the new edition which also includes a substantial 'Afterword' discussing scientists' reactions to the book.

1

A clean kill?: the role of Patriot in the Gulf War

In August 1990 Iraqi forces invaded Kuwait. The United States presented Iraq with an ultimatum – 'withdraw or face a military confrontation'. The Iraqi president, Saddam Hussein, responded by threatening to stage 'The Mother of All Battles'. Over the next four months the United States set about building up military strength in neighbouring Saudi Arabia with the intention of driving Saddam's army from Kuwait. Given Iraq's confrontational stance, this meant building a force capable of destroying all of Iraq's military resources.

Considering the scale of the imminent confrontation, and its distance from the American continent, the United States needed the backing of the United Nations and the military and political co-operation of many nations, notably Iraq's neighbours. A critical feature of this alliance was that a set of Arab states would side with the Western powers' attack on a fellow Arab state. As the old saying goes, 'my enemy's enemy is my friend', and at that time all the Arab states except Egypt had an enemy in common – Israel. On the other hand, America was Israel's staunchest ally, while Iraq was viewed as an important player in the confrontation with Israel. Thus the political alignment that the US needed to hold in place was continually in danger of collapse. It was crucial for American policy in respect of the forthcoming Gulf War that Israel did not take part in the conflict. Should Israel attack Iraq, creating circumstances in which the Arab states would be directly supporting Israel in its attack on an Arab ally in the Middle East conflict, it might become impossible for the other Arab states to continue to support America. Iraq's strategy was clear: they would try to bring Israel into the confrontation that had started with their invasion of Kuwait.

On 17 January 1991 the allies launched a massive air attack on Iraq that would last for five weeks. The war was over before the end of February following a devastating ground attack lasting four days. This is the setting for the argument that has raged about the effectiveness of the Patriot anti-missile missile.

On the first night of the air attack, Iraq fired six Scud missiles at Israel. The Scud was a Soviet-built missile extended in length and range by the Iraqis and known locally as the Al-Husayn. Thereafter Iraq launched many more Scuds at Israel, and at Saudi Arabia, especially at the American bases. On 25 February, a Scud hit an American barracks, killing twenty-eight and wounding ninety-eight military personnel. Otherwise, in spite of the fact that at least some Scuds landed and exploded, it was a failure in terms of its ability to damage men or materiel. As a propaganda and political weapon, however, it was, from the beginning of the war, potentially a potent force.

The Patriot was used in the Gulf War to combat the Scud. It was used first in Saudi Arabia and then it was rapidly deployed in Israel after the initial Scud landings. It may be that the military ineffectiveness of the Scud was due to the success of Patriot. It may be that irrespective of its military effectiveness, Patriot played an important role in keeping Israel out of the war; the fact is that Israel did not attack Iraq and the alliance held. During the course of the war the best information is that more than forty Scuds were directed at the allied forces and around forty at Israel. A total of forty-seven Scuds were challenged by 159 Patriots. The question is, how many Scuds did Patriot actually destroy?

WAR, SCIENCE, AND TECHNOLOGY

War is a confused and confusing business. Martin Van Creveld, the respected writer on military command, says on page 187 of his book *Command in War*, that war is 'the most confused and confusing of all human activities'. In the case of Patriot, technological fog and 'the fog of war' are found in the same place with quite extraordinary results. What we want to do is explain how it can be that it remains unclear whether the Patriot actually shot down any Scuds. Though

8

there are firm opinions on both sides, we still do not know whether the anti-missile missiles stopped Scuds from hitting Israel, stopped them from hitting Saudi Arabia, or failed to stop them at all.

The art of experimentation is to separate 'signal' from 'noise'. One would have imagined that one of the clearest 'signals' there could be would be the explosion of a ballistic missile warhead; one would think that this would be well out of the 'noise'. Either the Scuds were getting through and causing huge explosions, or the Patriots were destroying them and preventing the explosions – what could be a less ambiguous test of a technological system? It turns out that it was an extraordinarily poor test. The estimated efficacy of the Patriot missile in shooting down Scuds varies from around 100 per cent to around 0 per cent; some said every Scud warhead engaged was destroyed, some said not a single one was hit.

The story of estimates of Patriot's success starts at the beginning of 1991, during the war. Initially 100 per cent success was reported. The score steadily comes down to near zero by the time of a Congressional hearing in April 1992. The confidently stated figure initially moves to forty-two out of forty-five; to 90 per cent in Saudi Arabia and 50 per cent in Israel; to 80 per cent in Saudi Arabia and 50 per cent in Israel; to 60 per cent overall; to 25 per cent with confidence; to 9 per cent with complete confidence; to one missile destroyed in Saudi Arabia and maybe one in Israel. This is what happened as a result of ever more careful enquiries by US government agencies.

It is important not to misunderstand the figures at the lower end of the scale: they do not tell us how many Scuds were destroyed by Patriots; they are estimates of how many Scuds we can be confident were destroyed by Patriots. It may be that more Scuds were destroyed. But, if we are looking for a high degree of certainty, then our estimate has to remain low.

There are groups taking part in this debate with quite clear goals. In 1992, representatives of Raytheon, the manufacturer of the Patriot system, continued to claim that it had shot down most of the Scuds. On the other hand, Theodore Postol, the MIT academic who first drew public attention to doubts about optimistic claims for Patriot's success, continues to believe that he can prove that Patriot was an almost complete failure, and continues to press the case forward. Our interest and curiosity is sparked not by the actual success or otherwise of Patriot, but by the difficulty of settling the argument. We are not going to dwell on the interests, nor are we going to offer any conclusions as to whose evidence is biassed and why; we want to show only that the problem of measurement is hard to solve; we do not have, and cannot have, a clean scientific kill.

Was Patriot a success?

To say that the first casualty of war is truth is to miss the rather more important point that a principal weapon of war is lies. Disinformation confuses the enemy, while favourably biassed reports of success stiffen the resolve of one's own side and demoralise the opposition. It is, then, hardly surprising that, during the course of the war, Patriot was said to be a huge success. Not only was this important for the balance of morale, but it was vital that it was widely believed among Israel's populace that Saddam Hussein's forces were not being allowed to inflict damage on the Jewish State without opposition. It is fair to say that whether or not the politicians believed what they were saying, it would have been naive and unpatriotic of them to say anything other than that Patriot was a flamboyant success. It would be wrong to draw any conclusions for science and technology in general from wartime statements; wartime claims about the success of the missile reflect the demands of war rather than the demands of truth. Two weeks into the war, on 31 January, General Norman Schwarzkopf said, 'It's one hundred per cent so far. Of thirty-three engaged, there have been thirty-three destroyed.' A month into the war, on 15 February, President Bush said that forty-one out of forty-two of the missiles had been 'intercepted'.

After wars are over, the role of patriotic propaganda becomes less clear. Two weeks after the end of the war, on 13 March 1991, US Army officials told Congress that forty-five out of forty-seven Scuds had been intercepted by Patriots. Two months after the war's end, on 25 April 1991, the Vice-President of Raytheon, suppliers of the missile, said that Patriot had destroyed 90 per cent of the Scud missiles that were engaged over Saudi Arabia and 50 per cent of those engaged over Israel.

Any spokesperson for Raytheon has, of course, a clear interest in stressing the effectiveness of his corporation's product. Not only will a good reputation bolster future sales of the missile itself, but the proof of an anti-missile missile on the battlefield gives a boost to the prospect of whole new weapons systems. It gives credibility to the idea of a defence against the increasingly powerful nuclear, chemical, or biologically-armed missile arsenals belonging to so-called 'rogue states'.

Though the Israeli military seem to have been unimpressed by Patriot even during the war, US public confidence in its success was first dented by a professor from MIT's Defense and Arms Control Studies Program – Theodore Postol. In a fifty-page article, entitled 'Lessons of the Gulf War Experience with Patriot, published towards the end of 1991, Postol, drawing partly on Israeli sources, claimed that Patriot's performance 'resulted in what may have been an almost total failure' (p. 24). Much of what we will describe in this chapter has emerged as a result of Postol's critique.

Postol's work gave rise to a hearing before a Committee of Congress in April 1992 on 'Performance of the Patriot Missile in the Gulf War'. This provides a considerable source of unclassified information on the operation of Patriot and the arguments surrounding it. Much of the evidence given before the Congressional hearing by experts supports Postol's claims about the failure of the US Army to prove the success of Patriot, while disagreeing with his methods and his claims that he could prove it was a complete failure. In particular, Postol has used publicly available video-tapes to show that Patriot explosions did not disable Scuds because they did not occur in close proximity to the Scud warhead. The other experts claim that the frame-rate of commercially available video-tape is too slow to indicate the moment of explosion – Patriot would have travelled too far between frames. This bitter disagreement and its ramifications linger on at the time of writing (1997) but it is interesting that the enmity has not led to any disagreement about the lack of confidence in the high estimates of Patriot's success. Steven Hildreth, a 'Specialist in National Defense, Foreign Affairs and National Defense Division Congressional Research Service', disagrees with Postol's methods in the strongest terms. He stated, during the Congressional hearing: 'I think his case is worthless. I can say that.' Nevertheless, Hildreth claimed that using the US Army's own methodology, one can have confidence that only one Scud warhead was 'killed' by Patriot. This is not to say that only one warhead was killed, but that there is no really strong evidence for other successes.

Postol's work, and the Congressional hearing, attracted, in the summer of 1992, a twenty-five page riposte from Robert M. Stein, the Manager of Advanced Air Defense Programs for the Raytheon Company and a further response from Postol. Stein repeated an earlier comment made by Brigadier General Robert Drolet, US Army Program Executive Officer for Air Defense: 'In Saudi Arabia, Patriot successfully engaged over 80 per cent of the TBMs (tactical ballistic missiles), within its coverage zone [and] in Israel . . . Patriot successfully engaged over 50 per cent of the TBMs in the coverage zone.'

WHAT EVERYONE AGREES ABOUT PATRIOT

There are some features of the confrontation between Patriot and the Scud that everyone agrees about. Patriot was originally designed to shoot down aircraft rather than missiles. It had to be altered for use in the Gulf War and the design, development, manufacture, and deployment of the modified missile was carried through with heroic speed. There was a mistake in the missile software. This caused a timing error which meant that when the Scud fell on the Dhahran barracks, killing and wounding more than 200 people, no Patriot had been fired at it. This, however, does not detract from the major engineering feat that was involved in making the Patriot work at all in the Gulf War circumstances, nor is the software problem germane to other Scud engagements.

The version of the Scud missile used by the Iraqis during the Gulf War had been lengthened to extend its range. Its warhead seems to have been lightened and extra fuel tanks appear to have been added – perhaps cannibalised from other Scuds. Thus modified, the range was around 400 miles and it travelled some 40 per cent faster than had been anticipated by Patriot's developers.

The Al-Husayn rocket motor burns for about one-and-a-half minutes, taking the rocket to a height of around 35 miles, during which time it is guided along a pre-programmed trajectory. After this, all guidance ceases. The Al-Husayn then coasts for another 5 minutes, reaching a maximum height of around 100 miles, before

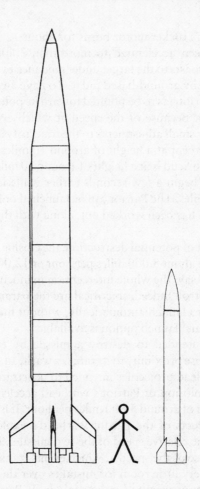

Figure 1.1.   Scud and Patriot (scale very approximate).

re-entering the atmosphere. It takes about another minute to impact with the ground. The rocket reaches a maximum speed of about 5,500 miles per hour before the atmosphere begins to slow it. At the point at which a Patriot might typically make an interception, perhaps 10 miles above the ground, the Al-Husayn might be travelling at about 4,400 miles per hour and would be less than 10 seconds from impact.

The Patriot's rocket motor burns for about 12 seconds, by which time it has been accelerated to more than 3,800 miles per hour. Thereafter it coasts to the target under guidance by its fins. The target is illuminated by ground-based radar. To have any chance of hitting the target, Patriot has to be pointed toward its potential impact point before launch. Because of the speed at which everything happens, only relatively small adjustments to the trajectory are possible during flight. To intercept at a height of around 10 miles, Patriot has to be launched when Scud is at a height of around 20 miles. Radar tracking of Scud must begin a few seconds earlier while it is at a height of, perhaps, 25 miles. The Patriot can be launched only once the path of the Scud's fall has been worked out, along with the potential impact point.

At the point of potential destruction, the closing speed of Scud and patriot will be about 8,000 miles per hour or 12,000 feet per second.

Needless to say, the whole interception manoeuvre must be under computer control. Indeed, so critical are the parameters that Patriot is usually set to launch automatically, without human intervention, though a manual launch option is available.

Patriot is intended to destroy a missile by exploding its own warhead in close proximity to its target's warhead. Patriot's warhead is jacketed in lead projectiles intended to penetrate vital parts of the Scud. The explosion of Patriot's warhead accelerates the lumps of lead to a speed of around 5,600 miles per hour. It is salutary to realise that the products of the Patriot warhead's explosion travel more slowly than the relative speed of the two missiles. Not only does this give one a sense of the speed at which events transpire, it also shows that there is very little room for mistakes over the time and place at which the Patriot warhead must explode. If Patriot's warhead explodes after it and Scud have crossed in the sky, there is no chance that the products of the explosion will catch up with the departing missile. Even if Patriot passes very close to its target, the time window for an effective detonation is very short. If Patriot misses the window by one-thousandth of a second, the relative positions of Scud and the interceptor will be different by about 12 feet. Instead of penetrating the explosive charge or fusing mechanism of the Scud, the lead projectiles will miss entirely, hit the empty fuel tanks, or cut the missile into pieces, leaving the explosive device intact.

To design an interceptor system that will identify incoming missiles, track them, determine their trajectory in a few seconds, fire a rocket at an estimated point of arrival, guide the interceptor to that point allowing it to make last second adjustments, and explode a warhead as it approaches the incoming missile, all at a closing speed of 8,000 miles per hour, is a remarkable achievement. It is still more remarkable when it is remembered that Patriot was originally designed to do something else. But all this achievement will be to no avail unless the warhead explodes at exactly the right time in exactly the right place. What might look like a perfect interception from the ground may have done nothing but damage the no longer essential parts of the incoming rocket's fuselage.

In Chapter 3 we distinguish 'demonstrations' from experiments and tests. War, because of its confusion and the extreme tension and terror under which soldiers operate, is not conducive to demonstrations. In war, most of the time, most things are going in unexpected directions. The Iraqi version of the Scud missile demonstrates this principle to perfection. Because the Al-Husayn had been lengthened and the warhead lightened, the Scud's aerodynamic and mechanical properties had changed. The unintended outcome was that Al-Husayn could re-enter the atmosphere after its long coasting flight in space in a variety of orientations, wobbling more or less depending on a variety of factors. Consequently it had a tendency to break into pieces during its descent through the atmosphere; the break-up could take place at different heights depending on the orientation and the wobble. The unevenly shaped pieces would then spiral to earth or take a still more random path depending on their shape and how they were affected by the airflow over them. The Iraqis had inadvertently designed a missile with decoy warheads capable of making the equivalent of evasive manoeuvres!

If the Scud broke up early in its flight the radar would 'see' what appeared to be two or three objects hurtling into the atmosphere instead of one. The Patriot system would automatically launch rockets to intercept all of these pieces, and in the early days Patriot stocks fell alarmingly. If the Scud broke up late in its flight the single Patriot, or pair of Patriots, launched to destroy one target would, during the course of the flight, suddenly be presented with several targets instead of one.

Thus, not only did Patriot have to destroy missile warheads when it had been designed to shoot down airplanes, it had to destroy a missile which behaved as though it had been built, by a very sophisticated design team, so as to avoid interception.

Two lessons can be taken from Patriot's performance in the face of the Iraqi Scuds' unexpected 'design feature'. One might say that Patriot worked even better than anticipated because it had to deal with unanticipated behaviour of its target. One might, on the other hand, draw the conclusion that anti-missile defensive systems will always be fighting the last war and evasive measures that cannot be anticipated will always defeat the defensive umbrella. Scud illustrates that these things will happen even when the missiles are built and launched by an unsophisticated military power; they give us pause for thought about what a sophisticated offensive force might do.

Later in the war it was learned that the dense, heavy warhead could be distinguished from the 'decoys' because it was slowed less by the atmosphere, but waiting long enough before firing to allow the difference to become clear meant that Patriot could not be launched until the Scud had fallen to a relatively low altitude. It appears that the Israelis switched from automatic firing mode to manual mode when they found out how to distinguish warhead from separated fuel tank and other debris, thus preventing their Patriot stock from being exhausted in the first days of the conflict.

CRITERIA OF SUCCESS

The debate about the success of Patriot turns on two interwoven issues. There is the question of what counts as success, and there is the question of whether Patriot achieved success under each of the many possible definitions.

Robert Stein, Raytheon's representative, published comments in the summer of 1992 exemplifying some of the things which might count as success; he mentions neither interceptions nor explosions:

> Patriot's very credible performance and success can be measured by the events as they occured. The coalition did not falter. Israel did not have to mount offensive actions against Iraq, and was able to stay

out of the war. Widespread loss of civilian life was not inflicted. . . .
And in the end it was Saddam who sued for peace, not the coalition.

We can think of the potential criteria for success from the inside out,
as it were; we start from the moment of embrace of Patriot and Scud
and work our way to less proximate consequences. The list in Table
1.1 represents some possible outcomes of the deployment of Patriot.

In this list, 'dudded' means that the warhead fails to explode when
it hits the ground as a result of Patriot's effects; 'damaged' means that
the warhead explodes but with reduced force (as might happen if
part of the explosive charge were distorted or separated); 'diverted'
means that a warhead travelling towards an area where it might
inflict death and destruction was diverted to an unpopulated area;
'intercepted' means that the radar tracking and flight control of the
Patriot was as expected, that Patriot approached the Scud and its
warhead fired, but it was not certain that the Scud warhead was
damaged.

The relationship between the last seven criteria of success, which
we call 'indirect criteria of success', and the first fifteen, which we call
'direct criteria of success', is not a straightforward matter of cause
and effect. The mere deployment of Patriot, along with the broad-
casting and widespread acceptance of stories of its success, could be
enough to bring about results 15 to 21, even if Patriot actually failed
on the direct criteria. Patriot could have been said to have been a
great success, either in the war or in the subsequent peace (a success
for the US arms industry), even if it had not intercepted or damaged a
single Scud; what is crucial is that people believed and believe that it
did. Ted Postol, a long-term opponent of Star Wars spending, is, as
he sees it, trying to prevent Patriot gaining the sort of false reputation
which would justify further expenditure on an anti-missile technology which, he believes, is bound to fail. Postol, in other words, is
trying to prevent consequences 20 and 21 coming about because he
believes that, in reality, consequences 1, 2, 4, 5, 7 and 8 did not
happen.

Another feature of lists such as that in Table 1.1 is that success
according to one criterion can be presented in such a way that it is
easy to read it as success under another criterion. Thus, two out of
the four claims that were mentioned at the outset: President Bush's

Table 1.1 *Criteria of success for Patriot vs. Scud*

Direct criteria of success
1   All, or nearly all, Scud warheads dudded.
2   Most Scud warheads dudded.
3   Some Scud warheads dudded.
4   All, or nearly all, Scud warheads damaged.
5   Most Scud warheads damaged.
6   Some Scud warheads damaged.
7   All, or nearly all, Scud warheads diverted.
8   Most Scud warhead diverted.
9   Some Scud warheads diverted.
10 All, or nearly all, Scuds intercepted.
11 Most Scuds intercepted.
12 Some Scuds intercepted.
13 Coalition lives saved and property damage reduced.
14 Israeli lives saved and property damage reduced.

Indirect criteria of success
15 Morale of civilian populations boosted.
16 Israel kept out of the war.
17 'Coalition did not falter.'
18 'Saddam . . . sued for peace not coalition.'
19 Patriot sales increased.
20 New anti-tactical missile programme given credibility.
21 Strategic Defense Initiative ('Star Wars') revivified.

'forty-one out of forty-two', and the US Army's forty-five out of forty-seven' referred to 'interception' not destruction. Whether this was a deliberately sophisticated use of words intended to give one impression while making a far less significant claim, is not clear. Senator Conyers cross-examined the US Army's Brigadier General Robert Drolet on this point during the 1992 Congressional hearing. The testimony speaks for itself:

> *Conyers*: Well was he [President Bush] in error?
> *Drolet*: No, sir.
> *Conyers*: So he was correct when he said forty-one out of forty-two Scuds were intercepted?

Drolet: Yes, Sir.

Conyers: You have records to back that up?

Drolet: Intercepted?

Conyers: Yes, sir.

Drolet: Yes, sir. He did not say killed or destroyed. He said intercepted. That means that a Scud came in and a Patriot was fired. But he did not say and we did not say, nor did we ever say, that it meant all of the Scuds were killed.

Conyers: He didn't mean that they were killed? He meant intercepted, meaning what in military jargon?

Drolet: . . . He just means that a Patriot and a Scud crossed paths, their paths in the sky. It was engaged.

Conyers: They passed each other in the sky?

Drolet: Yes, sir.

The indirect criteria of success

Sales, anti-tactical missiles, and Star Wars

Of criteria 19 to 21, we have little to say. It is interesting that many commentators, not least Robert Stein, were anxious to separate the last criterion from the others. As one commentator, who was hostile to Postol's view, put it during the Congressional hearing:

> . . . it has become fashionable to say, 'if Patriot, therefore, SDI' . . . an extrapolation from Patriot success to the probable future success of interceptors intended to hit strategic ballistic missiles armed with nuclear weapons.
>
> This is logically an absurd statement to make . . . A strategic defense system I would suggest, will probably get it right some time around the second or third nuclear war.
>
> *(Statement of Peter D. Zimmerman, Visiting Senior Fellow, Arms Control and Verification Center for Strategic and International Studies. Congressional hearing pp. 154–5)*

The anxiety seems to be to separate the positive lessons of Patriot for the development of defences against *tactical* ballistic missiles, from the credibility-sapping *Strategic* Defense Initiative meant to develop systems to shoot down the much faster inter-continental ballistic missiles. What is more, it is important to argue that Patriot and its ilk do not lead to a full-scale anti-ballistic missile system because there is an international treaty banning such developments. Those who want to develop Patriot-like missiles have to decouple them from the banned programme, whereas Postol argues that developing one is a surreptitious way of developing the other. At this stage of the argument, the success of Patriot in destroying Scuds is almost irrelevant.

Only history will tell whether Patriot was a success on criteria 19 to 21, but current evidence suggests that in respect of criterion 20 the Patriot experience was certainly not negative; anti-tactical ballistic missile systems are being developed, though Raytheon has a smaller part in this programme than they hoped for.

The local political role of Patriot

Of criteria 15 to 18, the argument is even more difficult to analyse. The history of warfare is a notoriously difficult topic because it is so invested with the interests of nations, armies, regiments, and generals, while the events themselves take place under circumstances where record-keeping is of less priority than survival.

Consider criterion 16, Patriot's role in keeping Israel out of the war. The *Washington Post* reporter, Rick Atkinson, says, in his history of the Gulf War, *Crusade: The Untold Story of the Gulf War*, that the Israeli military soon came to believe that the Patriot was performing poorly and they referred to Schwarzkopf's optimistic press comments as 'the Patriot bullshit'. He says however, enjoying, one presumes, the usual reporter's licence with quotations marks:

> Yet even the Israelis recognised the political and military utility in lauding the missile. When an Israeli officer suggested publicly disclosing qualms about Patriot, Avraham Ben-Shoshan, the military attache in Washington, snapped, 'You shut up. This is the best weapon we've got against the Scuds because it's the only weapon. Why tell Saddam Hussein that it's not working.' (p. 278)

In contrast, General Sir Peter de la Billiere, the commander of British forces during the conflict, does not mention Patriot at all in his biographical discussion, *Looking for Trouble*. He claims that it was British SAS ground patrols, deployed into Western Iraq on 22 January, that put an end to the Scud threat to Israel:

> . . . [they] operated with such effectiveness that no further Scuds were launched at Israel after 26 January. Once again, experience

confirmed that the SAS had repeatedly demonstrated in Europe –
that no amount of electronic surveillance is as effective as a pair of
eyes on the ground.   (p. 411)

We would prefer not to venture into the minefield of competing
accounts that comprise military history except to note that other
accounts say that thirteen Scuds fell on Israel between 28 January
and 25 February, wounding 34 people and damaging around 1,500
apartments and 400 houses.

Criterion 15, we would venture, was almost certainly met. It must
boost the morale of a civilian population to know that your side is
shooting back; the very act of resistance is cheering irrespective of
how successful that resistance is. It is said that later in the war
civilians took up positions on rooftops to watch the Patriot versus
Scud battle. This must be better for morale than cowering in shelters.

Death and destruction

Turn now to criteria 13 and 14. Did Patriot save lives and prevent
damage to property? There is no direct evidence in Saudi Arabia
because we do not know what damage would have been caused and
how many lives would have been lost if Patriot had not been deployed. We know that many servicemen and women were killed
when a Scud hit a crowded barracks, and we know that Patriot was
not deployed on that occasion, but it is impossible to be certain that
deployment would have made a difference. Therefore, the fact that
the greatest loss of life occured on an occasion when Patriot malfunc-
tioned cannot be taken to show that functioning Patriots were effective. We will see this more clearly as we proceed.

The case of Israel is more interesting because we have a 'before and
after' comparison. Scuds began to fall on Israel on 17 January but
Patriot was not deployed until around 20 January. A number of
missiles missed the towns at which they were putatively aimed, but
Postol says that thirteen unopposed Scuds fell onto Tel Aviv and
Haifa before Patriot was deployed, damaging 2,698 apartments and
wounding 115 people. After Patriot began to be used, he says,
fourteen to seventeen Scuds were engaged over the two towns and
7,778 apartments were damaged with 168 people wounded and one
killed. Postol says that on the face of it, three times as much damage

Figure 1.2.   Trails of Patriot anti-missile missiles over Haifa, Israel.
One appears to dive into the ground.

was caused by roughly the same number of Scuds during the Patriot
defence period compared to the pre-Patriot period. He admits, however, that since the amount of damage to each building was not
assessed in detail it is hard to rely on these rather slim statistics. He
concludes, nevertheless, that there was no evidence to suggest that
the damage decreased in the time of Patriot. The number of injuries
seems to have increased, and Postol suggests that some of these may
have been caused by Patriot itself. With Patriot deployed there would
certainly be much more in the air that would, at best, come down
again at high speed, and at worst, come down intact, exploding on

contact. There is evidence that some Patriots became confused by radar signals reflected from buildings, and that others chased Scud fragments into the ground.

Stein, in his reply to Postol, says that in fact far more missiles were directed at the Israeli cities after the deployment of Patriot, and that since the damage inflicted on the apartments was superficial, there is 'clear evidence that Patriot reduced ground damage other than in the category of "broken windows"'. (p. 222) He goes on to compare the relatively low loss of life and minimal damage in Israel with the severe damage and loss of life caused by tactical ballistic missiles in Iran's war with Iraq. He also goes on to compare the events in Israel with 'the lives lost in the Dhahran barracks when no defense counter-ed the TBMs'. (p. 222)[1]

The direct criteria of success

Interception and diversion

There has been little heated debate about whether Patriot *intercepted* most of the Scuds it engaged since interception says nothing about destruction, damage, or diversion. Let us define an 'interception' as meaning that the Patriot passed through a point, or potentially passed through a point, near to which its detonation would have dudded or damaged the Scud warhead.

We can say with certainty that not every Patriot launched intercep-ted a Scud. Some, especially in the early days, intercepted fuel tanks and other debris; some probably intercepted buildings, and some probably intercepted the ground. But this would be normal in a wartime situation.

Dhahran aside, there is no reason not to suppose that in every case, or nearly every case, that a Scud track appeared to be aimed at a populated area and was 'engaged', that one or more Patriots inter-cepted it in the sense expressed by Brigadier General Drolet – 'a Patriot and a Scud crossed paths, their paths in the sky'. Interception in this sense can be determined by radar traces; it does not require that we know what happened to the Scud after the Patriot exploded. This reliability of interception would be a considerable engineering feat, sufficient to bolster hopes for a successful anti-tactical missile system in the future, whether the Scuds were damaged or not. Of course, this is merely a start: it would remain to be proved that the Patriots could do significant damage to their targets, something still hotly disputed by Postol and others.

As regards diversion, the Al-Husayn was a very inaccurate weapon; unopposed it could not be expected to land within two or three kilometres of a well-defined target. Even the rockets that were tracked coming in did not have a predictable landing point because of the random aerodynamic forces caused by the break-up. The Army says that Scuds usually landed within 2 or 3 kilometres of predicted impact points.

Thus, the rocket that did so much damage to the Dhahran barracks could not be said to have 'hit its target' though it is possible that had it been opposed it might have been diverted. On the other hand, it might have been a matter of chance alone that Patriot did not divert another warhead onto an occupied building. Given the level of control, 'diversion' could be at best a chancy business. In any case, a diversion is a very different thing in the case of conventional explosive as opposed to chemical, biological, or nuclear warheads. Chemical and biological warheads will be affected by the direction of the wind; nuclear warheads would need much more diversion than a conventionally armed missile if it is to make a difference.

Dudding and damaging

All the above said, it is, on the face of it, surprising that we do not know whether Patriot reduced the number of 500 kilogram explosions that would otherwise have taken place in crowded cities. Because of the tiny margin for error of the Patriot warhead detonation if it was to damage the Scud warhead, we cannot know the answer to this from direct observations of Patriot interceptions. We have, therefore, to work out what happened from the actual Scud explosions and observations of ground damage.

Observations of explosions and ground damage are doubly confounded. On the one hand, even if every Scud warhead was dudded this would not mean there would be no damage. Bits of missile can hit the ground and cause a great deal of damage even when they are not fully functional. Thus, heavy warheads travelling at more than 3,000 miles per hour hit with the force of an explosion; the energy of the impact can even cause a flash of light which looks like an explosion. Fuel tanks, containing residues of rocket propellant, actually do explode when they strike.

It seems also that some Patriot warheads exploded at ground level, either by mistaking buildings for targets or because they chased Scuds, or Scud fragments, into the ground. Postol's video-tapes provide good evidence of ground-level Patriot explosions. Here we have an interesting question: does a ground-level Patriot explosion count toward the success of Patriot or against it? The answer depends on the criteria of success. If criteria such as 13, 14 or 15 are held in mind – reduction of damage to lives and property, and bolstering the morale of civilian populations, then a Patriot inflicting damage on the country it is meant to defend is seriously bad news. On the other hand, if one adopts the accountant's mentality and concerns oneself solely with criteria such as 1 through 9, 19 and 20, which have to do with Patriot's actual effectiveness against the Scud and its significance for the development of better systems, then the more ground damage that can be shown to have been caused by Patriot, the less of that ground damage can have been caused by Scud, and the more effective Patriot might have been at damaging Scud. Around here the argument becomes a little dizzying.

Working in the other direction, a Scud might land and do no more damage than might be expected from its energy of impact and exploding fuel, and this might be counted as a success for Patriot while in fact the Scud would not have exploded anyway; the Scud warhead might have been a dud! At least one 'warhead' landed which contained nothing but a lump of concrete – it seems that the Iraqis were running out of missiles equipped with fully functioning warheads towards the end of hostilities. In any case, we do not know how efficient they were at launching fully functioning missiles in the first place. Because of allied control of the skies, the Iraqis could not use fixed sites, as they could in their war with Iran; they were forced to use mobile launchers under conditions of secrecy and darkness and we cannot know how many warheads were properly prepared.

To understand the effectiveness of Patriot under criteria 1 through 6 it was vital to investigate ground damage outside urban areas as well as inside them. Furthermore, it now seems that the ground damage surveys were not conducted until some time after the war. The survey was conducted by 'one engineer working in Saudi Arabia for 5 days in February 1991, and 19 days in March 1991 . . . it relied heavily on photographs and interviews with military personnel assigned to the Patriot units, and . . . site visits were always made days or weeks after an impact when craters had often been filled and missile debris removed.' (Congressional hearing, report from US General Accounting Office, p. 78.) The Army's first report of Patriot effectiveness was also vitiated because the report included information on only about one-third of the Saudi engagements, although the Project Manager's assessment cites it as a source for all engagements. Furthermore, the report assumed that Patriot destroyed Scud warheads in the air unless warhead damage was found on the ground, while some units did not even attempt to locate damage. (p. 86)

One can now understand why it was that, by the time of the Congressional hearing, firm estimates of the success of Patriot had been dramatically lowered. The Army was now down to a claim of 60 per cent success rate, though that included engagements in which there was a low confidence of success. The Army also provided a 25 per cent success rate claim but that included 16 per cent of cases where the Patriot came close to the Scud without strong evidence that it destroyed or damaged it. Representatives of the US General Accounting Office said:

> . . . there is no way to conclusively determine how many targets the Patriot killed or failed to kill.
>
> About 9 percent of the Patriot's Operation Desert Storm engagements are supported by the strongest evidence that an engagement resulted in a warhead kill – engagements during which observable evidence indicates a Scud was destroyed or disabled after a Patriot detonated close to the Scud. For example, the strongest evidence that a warhead kill occured would be provided by (1) a disabled Scud with Patriot fragments or fragment holes in its guidance and fuzing section or (2) radar data showing evidence of Scud debris in the air following a Patriot detonation.
>
> *(Congressional hearing, p. 108)*

Steven Hildreth, we may recall, considered that, using the Army's own criteria, one could be certain of only a single warhead kill.

REACHING TOWARD THE LABORATORY

The experts giving evidence to the Congressional hearing agreed that to be sure about how well Patriot succeeded the battlefield would have had to have been equipped more like a laboratory or test range. High-speed photography might have revealed the exact moment of explosion of Patriot's warheads and their relationship to that of the Scud, but high-speed photography is not an option on the battlefield: 'The Chief Engineer said high-speed photography cannot be used to collect data unless the trajectory of both the Patriot and its target are known in advance so that multiple cameras can be set up along their flight paths. He also said this was not possible during Operation Desert Storm because obviously the time and location of Scud launches were not known in advance.' (Congressional hearing, p. 108, GAO report.)

Another expert, Charles A. Zracket, whose qualifications are given before the Congressional Committee, as 'Scholar in Residence, Center for Science and International Affairs, Kennedy School of Government, Harvard University, and the past President and Chief Executive Officer of the Mitre Corporation', states that '. . . the uncertainty of determining Patriot's actual shot-by-shot perform-ance in the gulf war comes about from the lack of high speed, high resolution photography and digital data radar recordings of inter-cepts that could provide direct and valid scientific data'.

There is also a possibility that a transmitter could have been attached to each Patriot that would have sent a continual stream of data back to the ground about its movements. This is known as 'telemetry'. Early in the war it came to be believed that telemetry equipment had caused a Patriot to fail and, thereafter, it was not attached to any of the missiles based in Saudi Arabia. The Israelis do appear to have used such equipment on at least a few Patriots. The GAO said that the telemetry data could have shown whether the Patriot passed within a range where it had a high probability of destroying the Scud and had time to detonate before the Scud flew past the intercept point. (p. 109) This, of course, would not prove whether the Patriot actually did destroy the Scud but would have resolved some of the ambiguity about whether the explosion hap-pened at the intended time and intended place.

In the evidence of the experts one sees a yearning for science in place of messy wartime reports: 'direct and valid scientific data' could put an end to this untidy debate.

There is a nice parallel between the performance of Patriot itself, and the performance of the monitoring equipment. Knowing what we know now we could, perhaps, redesign Patriot to intercept and destroy many more of the Al-Husayns. Patriots would be deployed in good numbers and optimum locations with well-trained crews well before the start of the war. The crews would already understand closing speeds and the typical trajectories and flight patterns of the missile warhead and its fortuitous decoys; new software would have been written accordingly. One might speculate that multiple Patriots would be directed at each warhead, each given slightly different fusing parameters. The software timing fault would have been rectified and the equivalent of the Scud that fell on Dhahran would be attacked and, perhaps, destroyed. In this scenario, Patriot would inflict clean kill after clean kill. As always, if only we could fight the last war again we would do it so much better. But then, we would not so much be fighting a war as rehearsing a large-scale manoeuvre and putting on a demonstration within its confines.

In the same way, the experts, knowing what they know now, understand what would be needed to produce a clean 'scientific kill' – an exact and unambiguous account of Patriot's encounters with its targets. Just as military men dream of fighting a war in which there is never any shortage of information or supplies, while the enemy always does the expected, so experts have their dreams of scientific measurement in which signal is signal and noise follows the model given in the statistical textbooks. As the generals dream of manoeuvres, so the experts dream of the mythical model of science.

But even manoeuvres and demonstrations rarely go to plan. Little is learned from a manoeuvre in which nothing goes wrong and, as we will see in our Chapter 3, demonstrations go wrong too, and even when they go right it is not clear what their implications are. Even when we have unlimited access to laboratory conditions, the process of measurement does not fit the dream; that was the point of our earlier book – the first volume of the Golem series. The point is not to show that the understanding of Patriot was beset by the fog of war, but to see that this fog was just a dense version of the fog through which golem science always has to strain to see. If the vision ever clears so far that we can get our shot in and produce a clean kill, then, as in the military situation, it is not likely to happen until after the metaphorical third 'nuclear war'; and by then it is usually too late for the purpose we have in mind.

NOTE

1 It has to be said that much of Stein's defence of Patriot's performance turns on statements of authorities which refer to, but do not detail, classified information. Furthermore, he presses home the message that the use of modified Scuds in the Gulf War is a warning for the future that the threat of offensive actions will not deter dictators from firing ballistic missiles at their enemies and that it is important to develop defensive systems against these threats. In his response, such messages, along with references to the more general criteria of success that come lower down our list, outweigh detailed analysis of Patriot's performance.

2

The naked launch: assigning blame for the *Challenger* explosion

We always remember where we were when we first heard about a momentous event. Those over forty-five years old know what they were doing when they heard that John F. Kennedy had been assassinated. Similarly, anyone who was watching television remembers where they were at 11:38 a.m. Eastern Standard Time on 28 January, 1986 when the Space Shuttle *Challenger* exploded. The billowing cloud of white smoke laced with twirling loops made by the careering Solid Rocket Boosters proclaimed the death of seven astronauts and the end of the space programmme's 'can do' infallibility.

Unlike the inconclusive Warren Commission that inquired into Kennedy's death, the Presidential Commission chaired by William Rogers soon distributed blame. There was no ambivalence in their report. The cause of the accident was a circular seal made of rubber known as an O-ring. The *Challenger*'s Solid Rocket Boosters were made in segments, and the O-rings sealed the gap between them. A seal failed and the escaping exhaust gas became a blow torch which burned through a strut and started a sequence of events which led to the disaster.

The Commission also revealed that the shuttle had been launched at unprecedentedly low temperatures at the Cape. Richard Feynman the brilliant, homespun American physicist is often credited with the proof. At a press conference he used a piece of rubber O-ring and a glass of iced water to show the effect of cold on rubber. The rubber lost its resilience. Surely this obvious fact about rubber should have been known to NASA? Should they not have realized that cold, stiff rubber would not work properly as a seal? Worse, it emerged that

Figure 2.1. *Challenger* explosion.

engineers from the company responsible for building the Solid
Rocket Boosters, Morton Thiokol, had given a warning. At an
impromptu midnight teleconference the night before the launch, they
argued that the O-rings would not work in the bitter cold of that
Florida morning. The engineers, it transpired, had been overruled by
their own managers. In turn the managers felt threatened by a NASA
management who expected its contractor to maintain the launch
schedule.

In 1986 the production pressure on the shuttle was huge. A record
number of flights – fifteen – had been scheduled for the supposedly
cheap, efficient, and reusable space vehicle. Among the cargoes were
major scientific experiments including the Hubble Space Telescope.
The launch preceding *Challenger* had been the most delayed in
NASA's history with three schedule 'slips' and four launch pad
'scrubs'. The ill-fated *Challenger* had already been delayed four
times and the programme as a whole needed to achieve a better
performance to fit the fiscal constraints of NASA in the 1980s. In the

glory days of the Apollo moon landings the 'Right Stuff' had been matched by the right money, but no more.

The conventional wisdom was this: NASA managers succumbed to production pressures, proceeding with a launch they knew was risky in order to keep on schedule. The *Challenger* accident is usually presented as a moral lesson. We learn about the banality of evil; how noble aspirations can be undermined by an uncaring bureaucracy. Skimp, save and cut corners, give too much decision-making power to reckless managers and uncaring bureaucrats, ignore the pleas of your best scientists and engineers, and you will be punished.

The *Challenger* story has its victims, its evil-doers, and two heroes. Aboard the shuttle was a school teacher, Christa McAuliffe, who represented ordinary people; she was one of us. The evil-doers were managers, both at NASA and at Morton Thiokol. One hero was Richard Feynman, who needed only a couple of minutes and a glass of cold water to establish what NASA had failed to learn in fifteen years. The other hero was the whistle-blowing, Morton Thiokol engineer, Roger Boisjoly. On the eve of the launch, fearing a catastrophe, Boisjoly tried desperately to get his company to reverse the launch decision. After the Rogers Commission vindicated Boisjoly, he took up the fight against Morton Thiokol and NASA in a billion-dollar law suit, alleging a cover-up. Boisjoly is the little guy taking on the Government and corporate America.

After the event it is easy to slot the heroes and villains into place. It is harder to imagine the pressures, dilemmas, and uncertainties facing the participants at the time that the fateful launch decision was made. We have to journey backwards in time to recapture just what was known about the O-rings and their risks before the launch.

Before we start, note how hard it is to sustain the image of amoral, calculating managers causing the crash. NASA managers and bureaucrats were under pressure to get the *Challenger* launched; yes, they knew as well as the next person that the endless delays put in jeopardy the image of the shuttle as a reusable and efficient space vehicle; and, yes, it would be a publicity coup for the hard-pressed space programme to get the Challenger launched in time for 'school teacher in space', Christa McAuliffe, to link up live with Ronald Reagan's forthcoming State of the Union address; but, why risk the space programme as a whole and their own futures for a matter of a few hours of scheduling? As George Hardy, then NASA's Deputy Director, Science and Engineering Directorate, Marshall Space Flight Center, told the Presidential Commission:

> I would hope that simple logic would suggest that no one in their right mind would knowingly accept increased flight risk for a few hours of schedule.
>
> *(Quoted in Vaughan, p. 49)*

Safety has to come high in any manager's calculation of risks and benefits however selfish and amoral their intentions.

New research on the *Challenger* launch decision by Diane Vaughan shows that the dangers of the O-rings were not ignored because of economic or political pressures. They were ignored because the consensus of the engineers and managers who took part in the fateful teleconference was that, based on the engineering data and their past safety practices, there was no clear reason not to launch that night. With twenty-twenty hindsight we can see they were wrong, but on the night in question the decision they reached was reasonable in the light of the available technical expertise.

O-RING JOINTS

A Solid Rocket Booster (SRB) works by burning solid fuel and oxygen. The shuttle SRBs are one hundred and forty nine feet tall, just shorter than the Statue of Liberty. They burn ten tons of fuel per second, which helps 'boost' the shuttle off the launch pad. Huge pressures of hot gas build up inside each booster, only relieved as the exhaust rushes out of the fireproof nozzle, providing lift. This exhaust gas jet is a potent force quite capable of melting metal. It is vital that it does not escape from anywhere except the place designed for it.

It is much easier to load solid fuel propellant into a booster which can be broken down into sections. Each booster on the shuttle (Figure 2.2) consists of four large cylindrical sections plus nose and nozzle. The sections are built at Thiokol's base in Utah, transported separately, and joined together at Kennedy. The joints between each section have to be specially designed to withstand the enormous

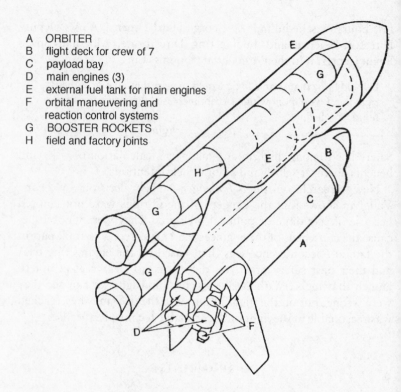

A   ORBITER
B   flight deck for crew of 7
C   payload bay
D   main engines (3)
E   external fuel tank for main engines
F   orbital maneuvering and
    reaction control systems
G   BOOSTER ROCKETS
H   field and factory joints

Figure 2.2.   Space Shuttle components.

pressures inside. During the first fraction of a second of launch, when the boosters first explode into life, each cylindrical section barrels outward causing the joints to bend out of true – a phenomenon known as 'joint rotation' (Figure 2.3). The barrelling outwards occurs because the joints with their supporting collar are much stiffer than the metal rocket casing above and below them.

The joints used are known as 'tang and clevis' joints. The cylindrical sections of the booster stack one on top of the other with the bottom lip of each section (the tang) fitting snugly inside the top of the section below in a special pocket just under four inches long (the clevis) (see Figure 2.3). Each joint is wrapped by a steel collar and fastened by 177 steel pins. The joints are sealed by two rubber O-rings nestled in grooves specially cut on the inside of the clevis in

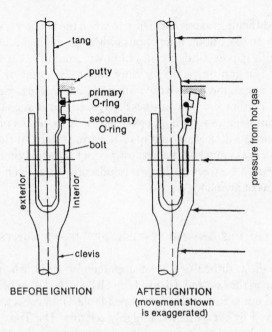

Figure 2.3.   Joint rotation.

order to prevent hot gases escaping during the fraction of a second of joint rotation.

An O-ring looks like thirty-eight feet of licorice, $\frac{1}{4}$ inch in diameter and formed into a loop. When the booster segments are stacked the O-rings are compressed (O-ring 'squeeze') to seal the tiny gap. A special putty based on asbestos protects the O-rings from the hot propellant gases. To test the joints before launch a leak check port is located between the O-rings. It is vital to make sure that the rings have seated correctly after the booster sections have been stacked. Air is blown between the two rings and the pressure is measured to check for leaks. Ironically this pressure test was later found to endanger safety because the air produced blow holes in the putty which the hot gases would seek out and penetrate, eventually reaching the O-ring seals and abrading them.

The joint between sections, like many components on the shuttle, is a lot more complicated than it looks. Even the question: 'What is the exact size of the gap between each section during launch?' turns out to be difficult to ascertain. The gap when the booster is stationary is known to be about four thousandths of an inch. But during ignition this gap expands by an unknown amount because of joint rotation. The gap opens up for only about six-tenths of a second. This may not sound like a very long time, but it is, as one engineer stated 'a lifetime in this business'. (Quoted in Vaughan p. 40.) Finding out what exactly happens during this six-tenths of a second takes an extraordinary amount of engineering effort. At the time the *Challenger* was launched the best estimates for the maximum size of the gap varied between forty-two hundredths of an inch and sixty hundredths of an inch.

THE DESIGN AND TESTING OF THE SRB JOINTS

None of these difficulties were apparent in 1973 when Morton Thiokol won the contract to build the SRBs. The design of the joints was based upon that of the very dependable Titan rocket. The joint was believed by everyone to be highly reliable. The Titan had only one O-ring. For added safety on the shuttle a secondary O-ring was added as back-up for the primary seal. The secondary seal provided 'redundancy'. NASA aimed for redundancy with as many components as possible, and particularly with ones where a failure would lead to losing the whole mission. Obviously not all components can have redundancy; if the shuttle lost a wing this would be a catastrophic failure, but no-one suggested that the shuttle should have a back-up wing.

Both NASA and its contractor, Morton Thiokol, took responsibility for the design and testing of the boosters and joints and seals. The tests were often carried out in parallel at Utah, where Thiokol was located, and at NASA's rocket engineering centre – the Marshall Center, Huntsville, Texas. Marshall was the jewel in NASA's crown. Its first director was the legendary German rocket scientist Wernher von Braun. He had established Marshall as the centre of technical excellence upon which the success of the Apollo programme was based. With this proud history, Marshall considered its expertise and facilities to be superior to anything Morton Thiokol could offer. The engineers at Marshall had a reputation for being conservative and rigorous; they saw it as their job to keep the contractor honest by trying to 'shoot down' their data and their analyses. At Thiokol the Marshall people became known as the 'bad news guys'. Thiokol's engineers favoured more practical options and were known to get 'defensive about their design', especially when under attack from Marshall. The differing attitudes of the two groups sometimes made for tense negotiations and led to some long-running disputes.

Both groups, of course, used the best science and engineering available to them. If there were technical conflicts they would try and resolve them until they both had the same results confirmed by multiple methods, multiple tests, and the most rigorous engineering analysis. But, as we shall see, rigorous engineering science, like golem science, is not always neat and tidy.

Early on the problem of joint rotation was recognized by both sets of engineers, but they differed in their estimates of its significance. Thiokol engineers calculated that, on ignition, the joint would close. Marshall engineers did not agree; their calculations showed that the joint would momentarily open. The consequences of the joint opening would be two-fold: (1) the compression or 'squeeze' on the O-ring would be reduced and make it a less reliable seal; (2) an O-ring could become unseated. Ironically in *this* disagreement Thiokol and NASA engineers took exactly the opposite positions they would take on the night of the *Challenger* launch. In this case it was NASA who expressed reservations about the joint while Thiokol was confident the joint would perform as the design predicted.

THE HYDROBURST TEST

To try and resolve matters a test was devised. The hydroburst test shoots pressurized water at the joint to simulate the pressures encountered at launch. It was conducted at Thiokol, in September 1977, where a joint was run through twenty pressure cycles – each cycle simulating the pressure at launch. The results indicated that NASA was right and that the joints could open for a short duration during ignition, blowing the seals out. Thiokol agreed with Marshall that the joint opened but they did not think that this threatened failure because the test was unrealistic. They had two grounds for challenging how well the test simulated an actual flight. Firstly, in a real flight, an O-ring would experience ignition pressure only once, not twenty times, and the data showed that the rings had worked perfectly for the first eight out of the twenty trials. Secondly, the test was carried out with the booster lying on its side. In use the booster would be vertical. They blamed the leaky joints on distortions in the horizontal rocket produced by gravity. Thiokol were so confident after this first test that they saw no need for any further tests. Marshall engineer Leon Ray, however, disagreed and insisted on more.

Similarity and difference

The problem the engineers are facing here is a variant of one raised by the philosopher, Ludwig Wittgenstein. Whether two things are similar or different, Wittgenstein noted, always involves a human judgement. We make such judgements routinely in our everyday lives. Things appear similar or different depending on the context of use. When, for instance, we see the familiar face of a friend from varied angles and under varied lighting conditions we have no trouble in saying *in the context of everyday recognition* that the face we see is always the 'same' face. In other words, we routinely treat this as a matter of *similarity*. If the context is, however, that of fashion photography, then crucial *differences* in the facial features emphasized in each photograph under varied lighting conditions or camera angles might become important. In this context we might routinely treat this as a matter of *difference*.

The same considerations apply to technological testing. Because most tests only simulate how the technology will be used in practice, the crucial question in judging test outcomes becomes: how similar is the test to the actual use? Morton Thiokol's interpretation of the hydroburst test was that the twenty ignition cycles test was sufficiently *different* to the actual use of the shuttle that it was not a good guide to what would happen during a real launch. NASA's position, on the other hand, was that the tests were *similar* to what might occur in a real launch. Therefore, for NASA, the tests were valid indications of problems with the seals.

As mentioned above, NASA–Marshall was well known for its conservative engineering philosophy. Thiokol engineers felt it had to be balanced with the need to produce a practical design:

You take the worst, worst, worst, worst, worst case, and that's what you have to design for. And that's not practical. There are a number of things that go into making up whether the joint seals or doesn't seal or how much O-ring squeeze you have, and they took the max and the min which would give you the worst particular case . . . all those worsts were all put together, and they said you've got to design so that you can withstand all of that on initial pressurization, and you just can't do that or else you couldn't put the part together.

(Quoted in Vaughan, p. 99)

With these underlying differences in design philosophy, and the interpretative loophole provided by the need to make similarity or difference judgements, it is not surprising that tests alone could not settle the performance of the joint.

MORE TESTING

More tests on the joints were carried out. These new tests, however, only made matters worse:

Continued tests to resolve the disagreement simply added to it. Marshall and Thiokol engineers disagreed about the size of the gap that resulted when the joint rotated. Although joint rotation was not necessarily a problem, the size of the gap mattered: if a gap were sufficiently large, it could affect the sealing capability of the rings. But tests to determine gap size continued to produce conflicting results . . . Both sides ran test after test, but the disagreement went unresolved.

(Vaughan, p. 100–101)

One of these tests, known as the Structural Test Article, simulated the pressure loads expected on the cylindrical sections of the shuttle during launch. Electrical devices were used to measure the amount of joint rotation. The results indicated that the problem was even worse than Marshall had first suspected. The size of the gap was sufficient for both O-rings to be out of position during ignition. The primary would be blown out of its groove as the pressure impacted on it, but the secondary would be left 'floating' in its groove, unable to seal should the primary fail later.

Thiokol again challenged Marshall's interpretation. In their view there was a problem not with joint rotation but rather with the electrical devices used to measure joint rotation. They maintained that these electrically generated measurements were 'off the charts' compared to their own (for them, obviously better) physical measurements of rotation which they had carried out during the test. They concluded that the calibration of the electrical instruments must have been wrong. And, since their own physical measurements indicated a smaller gap, the secondary would actually be in a position to seal.

Once again we are encountering the ambiguity of test results. Now the issue is another aspect of the human practices at the heart of golem science. The issue is that of *experimenter's regress* – a term introduced in *The Golem*. This refers to the catch-22 in research frontier experiments where the outcome is contested. The 'correct' outcome can only be achieved if the experiments or tests in question have been performed competently, but a competent experiment can only be judged by its outcome.

The logic of the situation is like this: what is the correct outcome of the test? Is there a large or small gap? The correct outcome depends on whether there is a large or small gap to detect. To find this out we must design a good test and have a look. But we don't know what a good test is until we have tried it and found that it gives the correct result. But we don't know what the correct result is until we have built a good test . . . and so on, potentially *ad infinitum*.

For NASA, the good test is the one with electrical data which supports their view that the gap is large enough to cause problems with the seals. For Morton Thiokol the good test is the one based on the physical measurements which supports their view that the gap is small and the seals will function as expected. Confidence for both also came from belief that their measuring equipment was the 'most scientific' and therefore produced the 'most accurate' and 'best' result.

Make sure the thing's going to work

Usually the experimenter's regress is soon broken by bringing in other considerations. If the two sets of engineers could not agree over the gap then perhaps another party could provide the necessary certainty. A third party with the requisite skills and familiarity with the O-rings was called in – the O-ring manufacturers. Unfortunately, in this case they could not resolve the issue either. They confirmed that the gap size was larger than industry standards and noted that the O-ring was being asked to perform beyond its intended design. But they then passed the buck back to the engineers, calling for yet more tests to be carried out which 'more closely simulate actual conditions' (Vaughan, p. 103). As NASA's Leon Ray recounted:

> I made the presentation to their technical folks, and we told them, the joints are opening up, and here's how much. What do we do? We've got the hardware built, we're going to be flying before long, what do you think we ought to do? And those guys said, well, that's a tough one, a real tough one. I feel sorry for you guys, I don't know what you're going to do, but we will study it over and give you an answer . . . Later on they both wrote letters back to us and said, hey, you've got to go with what [hardware] you've got. Do enough tests to make sure the thing's going to work. Now that's their recommendation. Go on and do some tests, and go with what you've had. That's all we can tell you. So we did.
>
> *(Vaughan, p. 103)*

Although a redesign of the joint was contemplated by Ray, at this late stage it would probably have badly delayed the shuttle programme. Also, all participants, including Ray, felt that the joint they had could be made to work; they just weren't certain it would work in all circumstances.

It is wrong to set up standards of absolute certainty from which to criticize the engineers. The development of an unknown technology like the Space Shuttle is always going to be subject to risk and uncertainties. It was recognized by the working engineers that, in the end, the amount of risk was something which could not be known for sure. They would do their best to make sure it was 'going to work'. Marshall's Larry Wear puts the attitude well:

> Any airplane designer, automobile designer, rocket designer would say that [O-ring] seals have to seal. They would all agree on that. But to what degree do they have to seal? There are no perfect, zero-leak seals. All seals leak some. It's a rare seal that doesn't leak at all. So then you get into the realm of, 'What's a leaking seal?'. From one

technical industry to another, the severity of it and the degree that's permissible would change, you know, all within the same definition of seals . . . How much is acceptable? Well, that gets to be very subjective, as well as empirical. You've got to have some experience with the things to see what you can really live with . . .

(Vaughan, p. 115)

And it should be borne in mind that the O-ring seal was just one of many components on the shuttle over which there was uncertainty.

TESTING FOR WORST SCENARIOS

A new strategy to resolve the dispute over gap size now emerged. What mattered most was not the exact size of the gap but whether the O-rings would actually work. Attention was switched to testing the seals' efficiency when gap sizes were much larger than either NASA or Thiokol expected. This time agreement was reached and, despite violating industry guidelines, the primary was found to seal under conditions far more severe than anything expected during a launch. As a further test, the performance of the secondary seal was also examined. A primary seal was ground down to simulate seal erosion and was tested at ignition pressure. In these circumstances, the secondary was found to seal, ensuring that the joint had the desired redundancy.

The two groups next tried to make the joints as tight as possible, producing maximum O-ring 'squeeze'. They did this by making the O-rings larger, ensuring they were of the highest quality and by 'shimming' the joints – wedging thin wafers of metal inside the joints to make them still tighter.

It was at this point, in July 1981, that Roger Boisjoly joined the Thiokol team. As was mentioned above, the amount of O-ring squeeze obtained was below the industry standard of 15 per cent (the minimum amount by which the rubber should contract under the pressure of the joint). Leon Ray for NASA and Roger Boisjoly for Thiokol worked closely together to reach a value acceptable to both sides. As Arnie Thomson, Boisjoly's supervisor, reported:

He [Roger] had a real good experience, hard, tough, head-knocking experience at first with Leon [Ray], because both happen to be

vigorous people, and both of them, you know, stand for what they
believe, and after, oh, gee, I guess it would be seven or eight months
of difficult conversations, Roger and Leon came to an agreement on
. . . about 7.5 percent squeeze [initial compression] . . . that was
negotiated, you know, that we would never go below [that], and we
would in fact select hardware [shims] so that that would not happen.
(Quoted in Vaughan, p. 104)

Ray, on behalf of NASA, was reported to be happy with the joint.
When the joint was tested he is quoted as saying:

We find that it works great, it works great. You can't tell any
difference. You don't leak at $7\frac{1}{2}$ percent.
(Quoted in Vaughan, p. 104)

Ray went on to summarize his views:

We had faith in the tests. The data said the primary would always
push into the joint and seal. The Titan had flown all those years with
only one O-ring. And if we didn't have a primary seal in a worst case
scenario, we had faith in the secondary.
(Quoted in Vaughan, p. 105)

Both engineering communities acknowledged that the joint did not
work as per design, but they felt that it worked sufficiently well that
it was an acceptable risk. Both groups now agreed that the primary
would seal, and both asserted that the joint had redundancy (i.e. the
secondary would act as a backup). But they had rather different
understandings of redundancy. NASA still believed there was a
larger gap than Thiokol believed. NASA felt that there would be
redundancy at initial ignition but not in the WOW (Worst on Worse)
situation such as a primary O-ring failing late in ignition when the
secondary O-ring might not be in position to seal. Thiokol, with their
smaller gap measurements, felt that there would be redundancy at all
times as the secondary would always be seated in the groove ready to
seal. As Boisjoly remarked:

In all honesty, the engineering people, namely Leon Ray . . . and
myself, always had a running battle in the Flight Readiness Reviews
because I would use the 0.042" [gap size] and they were telling me I
was using a number too low, and I would retort back and say no, the

horizontal number [Marshall's 0.060" gap size] doesn't apply
because we really don't fly in a horizontal position.

(Quoted in Vaughan, p. 106)

The Thiokol and Marshall designation of the joint as 'an acceptable
risk' was now officially endorsed by the NASA bureaucracy and the
shuttle was made ready for its first flight. The joints had passed
several formal bureaucratic risk evaluations as part of the process of
certification.

First flight of the shuttle

Now the ultimate test of the SRB joints could take place – the first
flight. Unlike in all previous tests, the gap between the *test* and the
actual technology in use could be finally closed.

On 12 April 1981, the first Space Shuttle was launched. Two days
later, after orbiting the earth thirty-six times, the shuttle landed
safely at Edwards Air Force Base. The two SRBs were recovered from
the ocean and disassembled. No anomalies were found in the joints;
they had behaved exactly as the Marshall and Thiokol engineers had
predicted.

1981–1985 EROSION AND BLOW-BY BECOME ACCEPTED AND EXPECTED

The second flight of the shuttle took place in November 1981.
Thiokol sent a team of engineers and a photographer to Kennedy to
inspect the recovered boosters. The manager of Thiokol's operations
at Kennedy, Jack Buchanan, was one of the first to notice something
had gone awry:

At first we didn't know what we were looking at, so we got it to the
lab. They were very surprised at what we had found

(Quoted in Vaughan, p. 120)

What they were looking at was one eroded primary O-ring. Hot
gases had burnt through the rubber by about five hundredths of an
inch. This may not sound like very much, and only one of the sixteen
joints had been affected, but O-ring erosion had never occurred on

Titan, nor on any of the test firings of the shuttle engines, nor on the previous flight. It was the first time. The engineers immediately started investigating what had gone wrong.

An explanation was soon found. The putty protecting the eroded seal was found to have tiny holes in it. These allowed propellant gases to reach the seal itself. As a Marshall engineer explained:

> The putty was creating a localized high temperature jet which was drilling a hole right into the O-ring.
>
> *(Vaughan, p. 121)*

Different compositions of putty and different ways of applying it were tried to prevent the problem repeating itself.

The eroded seal did bring some good news: there had been erosion, but the primary had still sealed. Also tests on the eroded seal enabled some specific safety margins to be drawn up. Small pieces were cut out of a primary O-ring to simulate almost twice as much erosion as had actually occurred. Even with this amount of erosion the joint was found to seal; this was tried at pressures three times greater than expected during ignition. As one Thiokol engineer commented:

> We didn't like that erosion, but we still had a couple of mitigating circumstances. First of all, it occurred very early on in firing . . . and if the primary O-ring was burnt right through that early . . . the secondary should be in a good position to catch it. In addition . . . we found out that even with major portions of the O-ring missing, once it gets up into that gap in a sealing position it is perfectly capable of sealing at high pressures . . .
>
> *(Vaughan, p. 122)*

With the new putty in place the next flight went according to plan. There was no erosion. This confirmed the belief amongst the engineers that they had the right explanation for the erosion.

Going operational

On July 4 1982, after the fourth flight had landed safely, President Reagan publicly announced that the shuttle was now 'operational'. Despite such proclamations, R&D continued apace on many components and glitches often occurred. For example, on the very flight which led Reagan to make his announcement, the Solid Rocket Boosters had mistakenly separated from their parachutes and had dropped into the ocean, never to be recovered. This deprived the engineers of an important set of data on the performance of the O-rings. The shuttle was certainly not an operational technology in the way that a commercial airplane is. Part of the shock caused by the *Challenger* accident comes from a mistaken image – an image NASA did nothing to discourage by flying US Congressmen and ordinary citizens in the vehicle. The shuttle always was, and will be for the foreseeable future, a high risk state-of-the art technology. Even today, with the joints redesigned after the *Challenger* accident and new safety procedures initiated at headquarters, the official risk is one catastrophic accident per hundred flights – astronomically greater than would be contemplated for any commercial vehicle.

More evidence of O-ring erosion was found in 1983 and 1984. The pattern was, however, frustratingly sporadic, usually only affecting one joint at a time, and the erosion could usually be explained away. In one case, the putty was again found to have had defects and in another case, the upping of the air pressure to check if the secondary was lodged in its groove before flight had induced blow holes in the putty. Tests and analysis continued and slowly the engineers thought they were getting a better understanding of how the joint behaved. It seemed that the hot gases only played on the rings for a very short period at the start of ignition and therefore only a limited amount of erosion occurred before the joint sealed. As the pressure across the joint equalized, the gas flow stopped and there was no more erosion. This led to the important notion that erosion was 'a self limiting factor'. The last two flights of 1984 had no erosion at all, seeming to confirm the engineers' belief that they were on top of the problem.

BLOW-BY

In 1985 the first cases of O-ring blow-by occurred. 'Blow-by' refers to hot ignition gases that blow past the primary during the split second before it seals. Blow-by is more serious than primary erosion because the hot gases threaten the secondary O-ring, jeopardizing joint redundancy. The first launch of 1985 experienced three consecutive nights of record low Florida temperatures. Blow-by on this flight reached a secondary O-ring.

After this instance of blow-by, Roger Boisjoly in particular felt there might be a link between low temperature and the damage. He had inspected the joint after its recovery and found grease in it. This grease had been burnt black and came from between the two O-rings. It was for him an indication that the cold may have affected the ability of the primary O-ring to seal. Boisjoly immediately started looking for systematic data on cold and O-ring performance. Part of the problem he faced was that all previous research had considered the effect of excessive heat on the O-rings. (The temperature in the main combustion chamber was 6,000 °F – above the melting point of iron.) Boisjoly had his 'observations' of the seal, but what he did not yet have were 'hard' data. The engineering culture at Marshall was such that for an argument to carry weight and be considered scientific, hard quantitative data were needed. Boisjoly set out to obtain these by initiating a test programme to investigate the effects of cold on the O-rings. This programme, however, was not urgent, ironically because the record low temperatures had been considered a fluke and unlikely to be repeated.

Although the engineers at Marshall and at Thiokol were alarmed about the first ever blow-by, they felt that they had a three-factor rationale for continuing to classify the joint as an acceptable risk. (1) The erosion was still within their basis of experience (it was less than the worst erosion experienced). (2) The amount of erosion that had occurred was within the safety margin established by testing a cutaway seal. (3) The phenomenon still seemed to be 'self-limiting'. Thus, despite the added worries about the temperature, Thiokol personnel (including Roger Boisjoly), concluded at the Flight Readiness Review for the next flight that '[it] could exhibit the same behavior. Condition is not desirable but is acceptable.' (Vaughan, p. 156)

On an April 1985 flight (launched at seasonally high temperatures) there was more blow-by when a primary O-ring burned completely through and hot gases, for the first time, eroded a secondary. This caused great concern, but a detailed analysis found an idiosyncratic cause for the blow-by. The erosion on the primary was so bad that it must have occurred in the first milliseconds of launch and this meant the primary could not have been seated properly. Since none of the other seals were damaged, that seal alone must have been impaired – a piece of hair or a piece of lint trapped unnoticed in the joint would have been sufficient. To prevent this happening again the leak test pressure was raised (an improperly seated seal would leak more). Again, even though the primary had failed, the secondary had worked, thus confirming redundancy. That the engineers seemed to understand the joint's performance was confirmed when primary O-ring erosion occurred on the very next flight, only to be explained by blow holes in the putty produced by the now greater pressure of the leak test!

Certainly the events of 1985, with blow-by encountered for the first time, caused mounting concern among the engineers. All sorts of new reviews, analyses and tests were ordered, including, as mentioned above, tests on O-ring resiliency at low temperatures. Despite this, all the engineers directly involved, including Roger Boisjoly, still considered the joint to be an acceptable risk.

THE *CHALLENGER* LAUNCH DECISION

We are now ready to understand what many accident investigators found incredible, how despite all the previous warnings and with Thiokol engineers pointing out on the eve of the launch the possible link between O-ring damage and low temperatures, the *Challenger* was still launched.

As we have seen, the joint was not perfect, but neither were a lot of other components on the shuttle. Also the relevant communities of engineers had, over the years, gained what they thought was an understanding of the joint's peculiarities; that hard-won understanding was not going to be given up easily for an untried new design which might have even more problems. Much of the misunderstanding over the *Challenger* accident has come, not only from twenty-twenty hindsight, but also, and more importantly for this book, from the mistaken view that engineering knowledge is certain knowledge. The expert engineers' own views of the risks versus the inexpert outside view can be seen in the following excerpt from testimony given to the Presidential Commission, where an FAA attorney, Ms

Trapnell, is interviewing Thiokol engineer, Mr Brinton:

> *Mr Brinton*: Making a change on a working system is a very serious step.
>
> *Ms Trapnell*: When you say working system, do you mean a system that works or do you mean a system that is required to function to meet the schedule?
>
> *Mr Brinton*: What I was trying to say is the colloquialism, 'If it ain't broke, don't fix it'.
>
> *Ms Trapnell*: Did you consider that system to be not broken?
>
> *Mr Brinton*: It was certainly working well. Our analyses indicated that it was a self limiting system. It was performing very satisfactorily . . .
>
> *Ms Trapnell*: Well, then, I guess I don't understand. You say on the one hand, that it was a self limiting situation . . . But on the other hand you say that the engineers were aware of the potential catastrophic result of burn-through.
>
> *Mr Brinton*: Well, let me put it this way. There are a number of things on any rocket motor, including the Space Shuttle, that can be catastrophic – a hole through the side, a lack of insulation. There are a number of things. One of those things is a leak through the O-rings. We had evaluated the damage that we had seen to the O-rings, and had ascertained to ours and I believe NASA's satisfaction that the damage we have seen was from a phenomenon that was self limiting and would not lead to a catastrophic failure.
>
> *Ms Trapnell*: . . . Does it surprise you to hear that one of Thiokol's own engineers [Boisjoly] believed that this O-ring situation could lead to a catastrophic failure and loss of life?
>
> *Mr Brinton*: I am perfectly aware of the front tire going out on my car going down the road can lead to that. I'm willing to take that risk. I didn't think that the risk here was any stronger than that one.
>
> *(Quoted in Vaughan, pp. 188–9)*

The problems with the joint were not, as many accident investigators surmise, suppressed or ignored by NASA, the engineers were actually all too well aware of the problems and the risks. They had lived with this joint and its problems over the years and they thought they had a pretty good understanding of its peculiarities. They knew the joint entailed a risk and, furthermore, a risk to the lives of the astronauts, but so did countless other components on the shuttle.

It is important to understand that the engineers and managers going into the crucial teleconference were not in a state of ignorance. They went in to the conference with all their accumulated experience and knowledge of the joint – any new information they were given was bound to be assessed in the context of this experience and knowledge and judged by the criteria they had always used.

The pre-launch teleconference

Let us follow events at Utah as the Thiokol engineers prepared for the crucial teleconference.

The main concern for Thiokol was that cold weather would reduce the O-ring resiliency. After a lengthy discussion and analysis, Thiokol decided that they would recommend 'no launch' unless the O-ring temperature was equal to or greater than 53 °F which was the calculated O-ring temperature on the previous coldest launch.

Thiokol felt that they did not have a strong technical position either way, but decided to act conservatively. The projected temperature of the O-rings (29 °F) at launch time was 24 °F below their lowest experience base. The flight with the lowest O-ring temperature had the worst blow-by. The O-ring squeeze would be lower; the grease would be more viscous; the O-ring actuation time – the time for the O-ring to be extruded into the seal gap – would be longer: consequently there must be doubts whether the primary and secondary O-rings would seal.

It was noted by the Thiokol engineers just before they went on air that there was a flaw in their technical presentation. A chart of Roger Boisjoly's, comparing two instances of blow-by, made the point that the January cold-temperature launch had the worse damage of the two. This chart, however, did not contain any temperature data. But another Thiokol chart revealed that the temperature for the second flight was 75 °F. By putting the two charts together, a huge gap in Thiokol's argument appeared. The two worst incidences of damage happened at both the highest and lowest temperatures!

The only way Thiokol could resolve the problem was to rely on Boisjoly's own observations that the blow-by really was a lot worse on the low-temperature launch.

The teleconference, with thirty four engineers and managers present, started at 8.15 p.m. (EST). The group did not divide neatly into engineers and managers since the structure of an engineering career means that everyone who was a manager had previously been a trained engineer. Thiokol presented all its charts and gave its rationale that the launch should be held back until the temperature reached 53 °F. The flaw in Thiokol's argument that temperature was correlated to damage was soon noticed. Boisjoly, referring to his visual inspection, was repeatedly asked if he could quantify his concerns, but he was not able to do so. Other inconsistencies appeared in the Thiokol presentation.

Larry Mulloy, the head of NASA's SRB management, who was at Kennedy for the launch, led the attack. His main point was that Thiokol had originated and up until now had always supported the three-factor rationale as to why the joint was an acceptable risk. And this was so even after the damage on the previous low-temperature launch was reported. Now they wanted to introduce temperature as a new factor, but O-ring blow-by could not be correlated with temperature according to their own data. Even the decrease in O-ring squeeze with temperature was not decisive. Although the resiliency was diminished, he argued that even at 20 °F the resiliency was positive and was greater than the minimum manufacturer's requirement. His conclusion was that the primary might be impaired in a worst case scenario, but that there was no evidence that the secondary would not seal.

All participants at the teleconference agreed that Thiokol had a weak engineering argument. The most controversial part was their recommendation of 'no launch' below 53 °F. This seemed to contradict their own data which had shown no blow-by at 30 °F. Procedurally it seemed odd to create a new criterion at the last moment – one which Thiokol had itself breached in the past. For example, for nineteen days in December and fourteen so far in January the ambient temperature at the Cape had been below the 53 °F limit. Indeed on the morning of the teleconference the ambient temperature was 37 °F at 8.00 a.m. and Thiokol had raised no concerns. It appeared to many participants that Thiokol's choice of 53 °F was somewhat arbitrary.

Mulloy made the point directly. Thiokol, in effect, were imposing a formal new criterion on the very eve of launch. This led to his infamous remark 'My God, Thiokol, when do you want me to launch, next April?' This comment was somewhat unfortunate in the light of what happened and was used against Mulloy as evidence that he put the flight schedule ahead of safety. Although Mulloy's statement was exaggerated for rhetorical effect it made clear his concern, that Thiokol were basing very serious conclusions and recommendations that affected the whole future of the shuttle on a 53 °F limit that he felt was not supported by the data.

Mulloy's views here were echoed by many of the participants at the teleconference. Marshall's Larry Wear said:

> . . . it was certainly a good, valid point, because the vehicle was designed and intended to be launched year-round. There is nothing in the criteria that says that this thing is limited to launching only on warm days. And that would be a serious change if you made it . . .
>
> *(Vaughan, p. 311)*

Another Marshall engineer, Bill Riehl, commented:

> . . . the implications of trying to live with 53 were incredible. And coming in the night before a launch and recommending something like that, on such a weak basis was just – I couldn't understand.
>
> *(Vaughan, p. 311)*

After Mulloy's 'April' comment the next to speak on the net was Mulloy's boss, George Hardy. Adding emphasis to Mulloy's argument he said that he was 'appalled' at the Thiokol recommendation of a 53 °F limit. Again he reiterated the weakness in Thiokol's engineering argument. He also added the well-remembered remark that, 'I will not agree to launch against the contractor's recommendation'. (p. 312)

Hardy's use of the word 'appalled' does seem to have carried weight with some of the participants. As one Thiokol engineer remarked:

> I have utmost respect for George Hardy. I absolutely do. I distinctly remember at that particular time, speaking purely for myself, that that surprised me . . . And I do think that the very word itself [appalled] had a significant impact on the people in the room. Everybody caught it. It indicated to me that Mr. Hardy felt very strongly that our arguments were not valid and that we ought to proceed with the firing.
>
> *(p. 312)*

Other participants, familiar with Hardy, and with the cut and thrust of these sorts of debates, felt that there was nothing unusual about Marshall's response. As Thiokol's Bill Macbeth said:

> No, it certainly wasn't out of character for George Hardy. George
> Hardy and Larry Mulloy had difference in language, but basically
> the same comment coming back, [they] were indicating to us that
> they didn't agree with our technical assessment because we had
> slanted it and had not been open to all the available information . . .
> I felt that what they were telling us is that they had remembered
> some of the other behavior and presentations that we had made and
> they didn't feel that we had really considered it carefully, that we
> had slanted our presentation. And I felt embarrassed and
> uncomfortable by that coming from a customer. I felt that as a
> technical manager I should have been smart enough to think of that,
> and I hadn't.
>
> (p. 313)

The Marshall engineers were certainly vigorous in their rebuttal. But this was quite normal. The two groups of engineers had been going at it hammer and tongs for years. What was new, however, was that this was the first time that the contractor had made a no-launch recommendation.

Thiokol requested a five-minute off-line caucus. Diane Vaughan reports that all participants who were asked why the caucus was called responded that it was because Thiokol's engineering analysis was so weak. NASA was apparently expecting Thiokol to come back with a well-supported recommendation not to launch, but with a lower and more reasonable threshold temperature.

Back in Utah, the five minutes turned into a half-hour discussion. Senior Vice President, Jerry Mason, chaired the discussion and started by reiterating the points Mulloy had made. Boisjoly and Arnie Thompson vigorously defended their position, going over the same data they had presented earlier. Mason put the counter arguments. The other engineers were mainly silent.

Finally, Mason said if engineering could not produce any new information it was time to make a management decision. None of the engineers responded to Mason's request for new information. Alarmed that their recommendation was about to be overturned,

Boisjoly and Thompson left their seats to make their arguments one last time. Boisjoly placed in front of Mason and his senior colleagues the two photographs of blow-by, which showed the difference in the amount of soot on the two launches. Sensing they were getting nowhere, the two engineers returned to their seats. Mason then polled his fellow senior managers: three voted to launch and one, Robert Lund, hesitated. To Lund, Mason said 'It's time to take off your engineering hat and put on your management hat.' Lund, too, voted to launch.

Subsequently, Mason's actions have been interpreted as replacing engineering concerns with a management rationale, where scheduling, the relationship between customer and client, and so on, were given priority. People who participated at the meeting, however, considered this to be a typical 'engineering management' decision made where there was an engineering disagreement. As Thiokol's Joe Kilminster explained:

> There was a perceived difference of opinion among the engineering people in the room, and when you have differences of opinion and you are asked for a single engineering opinion, then someone has to collect that information from both sides and made a judgement.
>
> *(Vaughan, p. 317)*

All the four senior Thiokol managers gave as their reason for changing their minds facts that they had not taken into account in their original technical recommendation. These were: no overwhelming correlation between blow-by and temperature; data showing that the O-rings had a large safety margin on erosion; and redundancy with the secondary.

The news of Thiokol's reversal was relayed as the teleconference went back on-line. After Thiokol's new recommendation and technical rationale had been read out, Hardy looked round the table at Marshall, and asked over the teleconference speakers if anyone had anything to add. People were either silent or said they had nothing to add. Finally the Shuttle Projects manager asked all parties on the teleconference whether there were any disagreements or any other comments concerning the Thiokol recommendation. No one said anything. The teleconference ended at 11.15 p.m. EST.

CONCLUSION

What has emerged from this re-examination of the shuttle launch is that the prevailing story of amoral managers, pressurized by launch schedules, overruling honest engineers, is too simple. There were long-running disagreements and uncertainties about the joint but the engineering consensus by the time of the teleconference was that it was an acceptable risk. Indeed, it was Thiokol's failure to meet the prevailing technical standards which led them to reverse their decision. They simply didn't have enough evidence to support a no-launch recommendation, particularly one that set a new low-temperature limit on the shuttle's operation that many considered unreasonable.

We are also now in a better position to evaluate another misconception – the one spread, perhaps inadvertently, by Richard Feynman: that NASA were ignorant of the effect of cold on O-rings. At the crucial teleconference this point was considered in some detail. NASA representatives had researched the problem extensively and talked about it directly with the O-ring manufacturers. They knew full-well that the O-ring resiliency would be impaired, but the effect was considered to be within their safety margins.

What the people who had to make the difficult decision about the shuttle launch faced was something they were rather familiar with, dissenting engineering opinions. One opinion won and another lost, they looked at all the evidence they could, used their best technical standards and came up with a recommendation.

Of course, with hindsight we now know that the decision they took was tragically wrong. But the certainty of hindsight should not be mistaken for the uncertainty of the golem at large. Without hindsight to help them the engineers were simply doing the best expert job possible in an uncertain world. We are reminded that a risk-free technology is impossible and that assessing the working of a technology and the risks attached to it are always inescapable matters of human judgement.

There is a lesson for NASA here. Historically it has chosen to shroud its space vehicle in a blanket of certainty. Why not reveal some of the spots and pimples, scars and wrinkles of untidy golem engineering? Maybe the public would come to value the shuttle as an extraordinary human achievement and also learn something about the inherent riskiness of all such ventures. Space exploration is thrilling enough without engineering mythologies.

Finally, the technical cause of the *Challenger* accident is to this day (1998) not absolutely certain. Cold temperature, erosion, and O-ring blow-by were a part, but other factors may have played a role: there were unprecedented and unpredicted wind shears on that tragic January day which may have shaken free the temporarily sealed joint. The current understanding of the size of the gap at rotation is that it is actually much smaller than either NASA's or Thiokol's best estimates. Ironically it is now believed that the excess squeeze from the much narrower gap was a contributory cause of the *Challenger* disaster. Golem technology continues its valiant progress.

All figures in Chapter 2 are reproductions of documents appearing in 'Report to the President of the Presidential Commission on the Space Shuttle Accident' (Washington, D.C.: Government Printing Office, 1986).

3

Crash !: nuclear fuel flasks and anti-misting kerosene on trial

The general public made the point, 'well that's all right, but we've got to take the word of you experts . . . for it – we're not going to believe that, we want to see you actually do it'. So well, now we've done it. . . . they ought to be [convinced]. I mean, I can't think of anything else. – If you're not convinced by this, . . . they're not going to be convinced by anything.

These words were uttered in 1984 by the late Sir Walter Marshall, chairman of Britain's then Central Electricity Generating Board (CEGB). The CEBG used the rail system to transport spent nuclear waste from its generating plants to its reprocessing plants. In spite of the fact that the fuel was contained in strong containers, or flasks, the public was not happy. The CEGB therefore arranged for a diesel train, travelling at a hundred miles per hour, to crash head-on into one of their flasks to show its integrity. Sir Walter's words were spoken to the cameras immediately following the spectacular crash, witnessed by millions of viewers either on live television or on the nation's televized news bulletins. Sir Walter was claiming that the test had shown that nuclear fuel flasks were safe. (The source from which Sir Walter's quotation was taken and of the basic details of the train crash is a video film produced by the CEGB Department of Information and Public Affairs entitled 'Operation Smash Hit'.)

In America, in the same year, a still more spectacular crash was staged. This time it was a full-sized passenger plane that was crashed into the desert. The plane was a radio-controlled Boeing 720 filled with dummy passengers. It was fuelled, not with ordinary kerosene but with a modified version, jellified to prevent it turning into an aerosol on impact. The jellification was brought about by mixing the kerosene with the same kind of additive as is used to thicken household paint. The fuel was meant to help prevent the horrendous fires that usually follow aircraft crashes and allow passengers who survived the impact itself to escape. The new fuel was known as Anti-Misting Kerosene, or AMK; it had been developed by the British firm ICI. The Federal Aviation Agency (FAA) arranged the demonstration. Unfortunately the impact was somewhat too spectacular; the plane caught fire and burned fiercely. A television documentary reported the event as follows.

> Before the experimental crash the FAA believed it had identified 32 accidents in America since 1960 in which AMK could have saved lives. But the awful irony is, the immediate result of the California test has been to set the cause back – maybe even to kill it off altogether. . . . those damning first impressions of the unexpected fireball in the California desert last December remain as strong as ever.

The two crashes seem to provide decisive answers to the questions asked of them. The fuel flask was not violated by the train crashing into it and the airplane burned up in spite of its innovative fuel. The contents of the fuel flask had been pressurized to 100 pounds per square inch before the impact, and it is reported that only 0.26 pounds per square inch pressure was lost as a result of the impact, demonstrating that it had remained intact. Photographs of the Boeing after the flames died left no doubt that no passenger could have survived the fire in the fuselage. Each of these experiments was witnessed indirectly – via television – by millions of people, and directly by hundreds or thousands. The train crash was watched 'on the spot', according to one newspaper, by '1,500 people including representatives of local authorities and objectors to nuclear power'. Though the laboratory experiments and astronomical observations we described in the first volume of the Golem series were untidy, ambiguous, and open to a variety of interpretations, and though there is some uncertainty about the effectiveness of the technologies described in the other chapters of this volume, these two demonstrations seem to have been decisive. Is it that the need for careful interpretation that we have described throughout the two volumes can be avoided if experiments are designed well and carried out before a sufficient number of eyewitnesses? Is it that the key to clarity is not the laboratory but the public demonstration?

TWO CRASHES: A SOLUTION TO TECHNOLOGICAL AMBIVALENCE?

According to the CEGB, the flasks used to transport spent nuclear fuel are made of steel. Their walls are about 14 inches thick and the lid is secured with 16 bolts, each capable of standing a strain of 150 tons. The flasks weigh 47 tons. For the crash, a flask and its supporting British Rail 'flatrol' waggon were overturned on a railway track so that the edge of the lid would be struck by a 'Type 46' diesel locomotive and its three coaches. Under the headline 'Fuel Flask Survives 100mph Impact', the *Sunday Times* of 18 July 1984 reported Sir Walter Marshall, chairman of the CEGB, describing it as 'the most pessimistic and horrendous crash we could arrange'. Sir Walter summed up the events for the TV cameras as described above (p. 57) stressing that no one should now remain unconvinced of the safety of the flasks.

NASA arranged that the Boeing 720 fuelled with AMK would land, with wheels up, just short of a 'runway' on which were mounted cutters – to rupture the fuel tanks and rip the wings – and various sources of ignition. 'Hundreds of spectators' watched. The *Sunday Times* (2 December 1984) said, 'A controlled crash to test an anti-flame additive went embarrassingly wrong yesterday when an empty Boeing 720 turned into a fireball . . .', and, 'A fireball erupted on impact'. Certainly a large fire started immediately, and the plane was eventually destroyed. In a television documentary called *The Real World* it was said that 'when the radio-controlled plane crashes into the desert, the British development team saw 17 years of work go up in flames'. The commentator described events as follows: ' . . . the left wing of the aircraft hits the ground first and swings the aircraft to the left and a fire starts. The senior American politicians cut short their visit to Edwards and the popular newspapers write AMK off as a failure.'

A nuclear fuel 'flask' is a hollow cube of metal, perhaps 10 feet high, with what look like cooling fins to the outside. In this case the

Figure 3.1.   Train crashes into nuclear fuel flask.

flask was painted bright yellow. The square lid is bolted to the top. The train approached from a distance. The locomotive was painted blue and it towed a string of brown-painted coaches. Seen on television, everything seemed to happen in slow motion, though this impression may be strengthened by the many subsequent slow motion replays. Because the senses are unprepared for such an impact, it is, in a strange way, quite difficult to see – one does not know what to anticipate and what details to look for. It is as though one is preparing to look into the sun – one almost screws up one's eyes in anticipation. The train hit the flask with a tremendous explosion of dust, smoke, and flame, pushing the flask before it. Parts of the engine flew in all directions. Once the dust had settled one could see the flask among the debris. A section of the fins had been gouged by the impact but, as the pressure test showed, apart from cosmetic damage, the flask was otherwise inviolate. A promotional video of the event, made by the CEGB, describes how fuel flasks are constructed and repeats the history of the test; it shows the crash, underscoring its significance.

With the airplane crash the sense of witnessing the, literally,

Figure 3.2.   Boeing 720 about to crash into obstacles in 'demonstration' of anti-misting kerosene.

inconceivable was even more pronounced. Here was a full-scale passenger plane, something of obvious high financial and technological value, something that we are used to thinking of as a symbol of exaggerated care and maintenance, shining silver against the crystalline blue sky, about to crash into the pale desert floor, made even more perilous by jagged cutters, designed to rip the wings apart, set into the mock runway. The airplane struck with a vicious tearing; parts flew off; slow motion showed a door ripped off and the immediate billowing up of a red-yellow flame which enveloped the aircraft. A distant shot showed a column of black smoke climbing high into the desert sky. Subsequent close-ups showed the burnt-out fuselage and the burnt remains of dummy passengers.

The author of this chapter (Collins) has tried his own experiment, showing videos, or sets of slides, of the train crash to various audiences, and asking them to point out any problems – any reason why we should refuse to believe the evidence of our senses and fail to accept Sir Walter Marshall's claim. While Sir Walter's presentation, with its used-car-salesman's brashness, tends to encourage cynicism, audiences have rarely been able to come up with any technical reason for drawing contrary conclusions. Seeing the film of the plane crash in its initial cut has the same effect – one cannot imagine anything other than that it shows that AMK failed. The contrast with the ambivalent scientific experiments and technological forays we have described elsewhere in these volumes seems stark. We need, however, a more sophisticated understanding: we need first to make some conceptual distinctions between experiments and demonstrations, and then we need to re-analyse these two incidents with the help of experts. The distinction between experiments and demonstrations was, of course, central to the argument of Chapter 1, on the Patriot missile.

EXPERIMENTS AND DEMONSTRATIONS

The Golem series deals with matters which are controversial. In so far as the terms have any meaning at all, making use of a commonplace technological artifact – driving in one's car, writing a letter on one's word-processor, storing wine or whiskey in a barrel – is a demonstration of the reliability and predictability of technology: these things work; about such things there is no controversy. And that is why they are politically uninteresting – the citizen need know nothing of how such things work to know how to act in a technological society. But each of these technologies may have a more politically significant controversial aspect: Are diesel cars more or less polluting than petrol cars? Is literacy being destroyed by the artificial aids within the modern word-processor? Are the trace chemicals leached out of smoked barrels harmful to health (we have not the slightest reason to think they are)? The treatment of the controversial aspects must be different to the uncontroversial aspects. The same is true of what we loosely refer to as experiments: one does not do *experiments* on the uncontroversial, one engages in *demonstrations*. Historically, as Steven Shapin argues, there has been a clear distinction between these two kinds of activity.

In mid to late seventeenth-century England there was a linguistic distinction . . . between 'trying' an experiment [and] 'showing' it . . . The trying of an experiment corresponds to research proper, getting the thing to work, possibly attended with uncertainty about what constitutes a working experiment. Showing is the display to others of a working experiment, what is commonly called demonstration. . . . I want to say that trying was an activity which in practice occured within relatively private spaces, while showing . . . [was an event] in relatively public space.

To move to the nineteenth century, Michael Faraday's mode of operation at the Royal Institution shows the same distinction between experiment and demonstration. Faraday practised the art of the experimenter in the basement of the Royal Institution but only brought the demonstration upstairs to the public lecture theatre once he had perfected it for each new effect. As David Gooding puts it:

> Faraday's *Diary* records the journeys of natural phenomena from their inception as personal, tentative results to their later objective status as demonstrable, natural facts. These journeys can be visualised as passages from the laboratory in the basement up to the lecture theatre on the first floor. This was a passage from a private space to a public space. (p. 108)

The hallmark of demonstrations is still preparation and rehearsal, whereas in the case of an experiment one may not even know what it means for it to work – the experiment has the capacity to surprise us. Demonstrations are designed to educate and convince once the exploration has been done and the discoveries have been made, confirmed, and universally agreed. Once we have reached this state, demonstrations have the power to convince because of the smoothness with which they can be performed. Indeed, the work of being a good demonstrator is not a matter of finding out unknown things, but of arranging a convincing performance.

Demonstrations are a little like 'displays' in which enhancements of visual effects are allowed in ways that would be thought of as cheating in an experiment proper. For example, in a well-known lecture on explosives, the energy of different mixtures of gases is revealed. If the mixtures are ignited in milk bottles they make increasingly loud bangs as more energetic reactions take place. When I watched such a lecture, a milk bottle containing acetylene and oxygen, the most powerful mixture, was placed in a metal cylinder while a wooden stool was set over the top of the whole arrangement apparently to prevent broken glass spreading too far. When the mixture exploded the seat of the wooden stool split dramatically into two pieces. Subsequent inspection [by the author] revealed that the stool was broken before the test and lightly patched to give a spectacular effect. In an experiment, that would be cheating, but in a display, no one would complain. A demonstration lies somewhere in the middle of this scale. Classroom demonstrations, the first bits of science we see, are a good case. Teachers often know that this or that 'experiment' will only work if the conditions are 'just so', but this information is not vouchsafed to the students.

The presentation of science to the general public nearly always follows the conventions of the demonstration or the display. If we are not specialists, we learn about science firstly through stage-managed demonstrations at school, and secondly through demonstrations-cum-displays on television. This is one of the means by which we think we learn that the characteristic feature of scientific tests is that they always have clear outcomes. A demonstration or display is something that is properly set before the lay public precisely because its appearance is meant to convey an unambiguous message to the senses, the message that we are told to take from it. But the significance of an experiment can be assessed only be experts. It is this feature of experiment that the Golem series is trying to make clear; some more recent television treatments of science are also helping to set the matter right. It is very important that demonstration and display on the one hand, and experiment on the other are not mistaken for one another.

One final and very simple point is worth making before we return to the story of the two public 'experiments'. The simple point is that it is not true that 'seeing is believing'. Were this true there would be no such profession as stage magic. When we see a stage magician repeat an apparently paranormal feat, even though we cannot see the means by which it was accomplished, we disbelieve the evidence of our senses; we know we have been tricked. The same applies in the case of the 'special effects' we see at the cinema, the case of some still photographs, the case of 'optical illusions' on the printed page and the case of every form of representation that we do not read at face value.

Contrasting, once again, the demonstration with the stage magician's feats, whether we believe what we see is not solely a matter of what we see but what we are told, either literally – as in the case of Sir Walter Marshall's comments to the television – or more subtly, through the setting, the dress, and the body language of the performers, and who is doing the telling; it is these that tell us how to read what we see.

THE CRASHES REANALYSED

The train crash revisited

The self-same manipulation of reality can be both an experiment and a demonstration. It can demonstrate one thing while being an experiment on another; it can fail in one aim while succeeding in the other. Both crashes were like this, but in different ways. The train crash was a superb demonstration of the strength and integrity of a nuclear fuel flask, but at the same time it was a mediocre experiment on the safety of rail as a means of moving radioactive materials. The plane crash was a failed demonstration of the safety of AMK but, as we will see, it was a very good experiment. There is nothing more that needs saying about the demonstration aspect of the train crash, but it is worth pointing out that Sir Walter, and the subsequent promotional material, encouraged us to read it as a successful experiment on nuclear waste transport policy. To find out why it was not such an experiment, we need to turn to the experts.

The conservationist pressure group 'Greenpeace' employs its own engineers. Their interpretation was somewhat different to that of Sir Walter Marshall. By their own admission, the CEGB were so sure that the flask would not break under the circumstances of the crash, that they could learn little of scientific value from it. They already understood quite a lot about the strength of flasks from previous private experiments and knew the circumstances under which they would not break. If Greenpeace are to be believed, CEGB also knew circumstances under which the flasks would break. Greenpeace did not question that this particular flask had survived the 100 m.p.h.

impact intact, but they suggested that even as a demonstration, the test showed little about the transport of nuclear fuel in 1984. They claimed that the majority of the flasks used at the time were of different specification: the test specimen was a single forging of 14-inch thick steel, whereas the majority of the CEGB's flasks were much thinner with a lead lining. Flasks of this specification, they suggested, had been easily broken by the CEGB in various preparatory experiments done in less public settings.

Furthermore, Greenpeace did not accept that the test involved the most catastrophic accident that could occur during transport of nuclear fuel. Greenpeace held the crash to be a far less severe test than could have been arranged; they believed that the CEGB, through their consulting engineers Ove Arup, were aware of a number of crash scenarios that could result in damage even to one of the stronger, 14-inch thick, flasks. These included the flask smashing into a bridge abutment at high speed after a crash, or falling onto rock from a high bridge. Thirdly, they believed that the design of the test carefully avoided a range of possible problems. The CEGB, said Greenpeace, had the information to stage things in this way because, while developing the demonstration, they had done extensive work on different types of locomotive, different types of accident, and so forth.[1]

Specifically, Greenpeace claimed that the 'Type 46' locomotive used in the test was of a softer-nosed variety than many other types in regular use. Thus, to some extent, the impact was absorbed in the nose of the engine. They claimed that the angle of impact of train to flask was not such as to maximise the damage. They claimed that the carriages of the train had been weighted in such a way as to minimize the likelihood that they would ride over the engine and cause a secondary impact on the flask. They claimed that because the wheels of the flatbed waggon used to carry the flask had been removed, there was no chance of them digging into the ground and holding the flask stationary in such a way as would increase the force of impact. (The film of the crash shows that the waggon and flask were lifted and flung through the air by the train, offering little resistance to its forward motion.) Finally they claimed that the waggon and flask were placed at the very end of the rail track so that, again, there was nothing in the way of sleepers (ties) or track to resist the movement of

Figure 3.3. Nuclear fuel flask being examined for damage after impact.

the flask as it was pushed along the ground – everything beyond the impact zone was smooth earth – there was nothing to hold the flask or penetrate.

We do not know whether the commentary of Greenpeace's engineers is more accurate than that of the CEGB and this is not our business. We cannot judge whether Greenpeace or the CEGB had it right. Perhaps the flask would have survived even if the wheels had been left on the waggon and the rails had been left in place. We do not have the expertise to be sure.

Given all this, the crash test seen on TV may have been spectacular and entertaining, but there are many ways in which a real experiment could have been a more demanding test of the waste transport policy. The test demonstrated what the CEGB already knew about the strength of flasks, but it did not show that they could not be damaged under any circumstances. Sir Walter Marshall, as represented on TV, encouraged the public to read the outcome not as demonstrating the CEGB's mastery over a small part of the physical universe, but as demonstrating that nuclear flasks were safe. The public, then, were put in a position to read a demonstration of one small thing as a proof of something rather more general.

The plane crash revisited

In the case of the plane crash, several experts were interviewed by the makers of a British television programme and it is this expertise that we tap.[2] This programme attempted to reinterpret the crash so as to support the idea that the test had demonstrated not the failure of AMK, but its effectiveness. The programme reassembled much of the film shot from the various angles, overlaying it with explanation and comment from representatives of the Federal Aviation Authority and Imperial Chemical Industries (ICI), the British firm that made the jellifying additive and developed AMK. We see, among other things, that television does not have to simplify.

In spite of the fire and the destruction of the airplane, the programme team claimed that the test was not such a failure for AMK as it appeared to be. We will use our demonstration/experiment dichotomy to describe the findings of the programme team, treating the demonstration aspects first and the experimental aspects second.

The demonstration should have resulted in no fire, or a small fire at most. The demonstration had been worked up by studying many collisions using old non-flight-worthy airplanes propelled at high speeds along test tracks. This series of tests had shown that AMK should work; it remained only to convince the public and the aviation pressure groups with a spectacular test. According to an ICI spokesman,[3] the Federal Aviation Agency, NASA and ICI were all convinced of the effectiveness of AMK before the test, and thought of the test as a demonstration. But the demonstration did not go exactly as planned; the ground-based pilot lost control of the radio-controlled plane. Instead of sliding along the runway with wheels down, the plane hit the ground early, one wing striking before the other. The plane slewed to one side, crashed into the obstacles and came to a rapid stop. A metal cutter designed to rip into the wings and fuel tanks entered one of the still spinning engines, stopping it instantly and causing a far greater release of energy than would normally be expected. This, in turn, caused the initial fireball which would not have happened if the demonstration had gone according to plan.

It might be argued that airplane crashes never go according to plan, but this is to miss the point that AMK had already been

Figure 3.4. Two identical sled-driven planes, fuelled with ordinary kerosene (top) and anti-misting kerosene (bottom), are crashed in tests prior to the public demonstration.

shown to be safer than ordinary kerosene under many crash circumstances. Most crashes would not involve an impact between a spinning turbine and a rigid metal object. If AMK could improve survival chances in most crashes that would be enough to merit its use; there was no need for it to be effective in every imaginable case. Therefore it was appropriate to demonstrate the effectiveness of AMK in a demonstration of a typical crash, rather than an extreme crash such as the one that actually happened. Bear in mind also that there is no point in exploring the effectiveness of AMK in crashes that are so violent that the passengers would not have survived the impact; this again makes it clear that it was a relatively mild crash that was intended, and appropriately intended. (Compare this with the nuclear fuel flask case, where even one accident involving the release of radioactivity would be unacceptable.) It is clear that what we have in the case of the plane crash is a demonstration that went wrong.

The very fact that the demonstration went wrong moved the event into unknown territory for which the scientists and technologists had not prepared; that is, the intended demonstration turned into an experiment. An ICI spokesman said, 'in [accidentally] introducing a number of unknown effects we were able to observe things that we had not seen before and were able to draw a number of useful conclusions from them.'[4] Appropriately interpreted, this experiment can be read as a success for AMK even under the most extreme of survivable conditions.

The initial fireball, horrendous though it appeared, was less severe than might have been expected given the nature of the crash. The fire swept over the fuselage, but did not penetrate into the passenger cabin or the cockpit. It is claimed that a proportion of the AMK which spilled as a result of the first impact did not burn; on the contrary, it splashed over the fuselage and cooled it.

This initial fire died out after a second or so. Because it did not penetrate the fuselage, this fire was survivable if terrifying. Passengers could have escaped after the initial fireball had died away. An airplane fuelled with ordinary kerosene involved in a crash of this sort would have burned out immediately, giving the passengers no chance.

Though the plane was eventually destroyed, the major damage was caused by fuel entering the plane through an open cargo door and through rents in the fuselage made by the cutters as the plane slewed to one side. This second fire, though it was not survivable, did not start until after the passengers, or at least some of them, would have had time to escape.

Furthermore, the test showed in another way that AMK is far less flammable than ordinary jet fuel. This can be seen because even after the second fire had destroyed much of the plane, there were still 9,000 gallons of unburnt fuel left in and around the plane. The programme showed film of this unburnt fuel being salvaged.

A spokesman for ICI says on the programme:

> Subsequently, with the analysis of what happened, those who have been technically involved actually believe it was a success – that . . . the aircraft had a smaller fire than had been expected if there had been jet-A [ordinary aircraft kerosene] – a benefit that we hadn't even expected at all in that . . . the anti-misting fuel helped to cool the aircraft and provide conditions within the aircraft which would have allowed people – some people – to escape.

Here television had a role to play in defining the test as a failure. The critical period is between the initial fireball and the subsequent extensive fire. It is during this period that passengers still able to move could have escaped from the aircraft. And yet the natural way of cutting this episode for television was to move from the first fireball to the smoke cloud in the desert: there is nothing spectacular about the long period in-between when nothing happens. One must remember that the direct evidence of the senses is not the same as television. Anything seen on television is controlled by the lens, the director, the editor and the commentators. It is they who control the conclusions that seem to follow from the 'direct evidence of the senses'. A respondent who was present at the NASA test – the scientist who developed the fuel additive – said that the meaning of the test was by no means obvious to those who were actually there.

> The observation points were about half a mile from the actual landing site and it took about an hour to assemble back at the technical centre. The instantaneous reaction of the people in my immediate vicinity and myself, when we saw that, after the fire-ball (larger than expected) died away (about 9 seconds), the fuselage appeared to be intact and unscathed, was success – a round of applause!

In the case of the plane crash we are extraordinarily lucky to have been able to watch a television reconstruction informed by experts who had an interest in revealing the ambiguities that are normally hidden.

Once more, we do not intend to press ICI's case against the interpretation of the local experts, but simply to show that as soon as the plane crash is seen as an experiment, rather than demonstration, enormous expertise is required to interpret it, and a variety of interpretations is possible. It is not our job to choose one interpretation

Figure 3.5.   What the television viewer saw: the Boeing 720 crash as photographed from the television screen.

over the others; it is not that we do not care – after all, human lives depend on the solution – but it is not our area of expertise.

IMAGINING WHAT MIGHT HAVE BEEN DONE

We can reinforce the distinction between demonstration and experiment by imagining how these things might have been done differently. Imagine how Greenpeace might have staged the demonstration if they had been in control. One would imagine that they would first experiment in private in order to discover the conditions under which minimum force *would* damage a flask. An appropriate demonstration might involve firmly positioned rails which would penetrate one of the earlier thin-walled flasks when it was struck by a hard-nosed locomotive going at maximum speed. Perhaps the accident would take place in a tunnel where the flask had initially been heated by a fierce fire. About six months after the CEGB's experiment (20 December 1984), a train of petrol tankers caught fire inside a tunnel producing temperatures well in excess of anything required during statutory tests of fuel flasks. It seems possible that a flask caught in that fire would have been badly damaged. At least the event revealed the possibility of a hazard not mentioned during the CEGB's crash test.

A simpler, but effective, Greenpeace demonstration could reveal what would happen *if* a nuclear flask were damaged, without worrying too much exactly how the damage might be caused (or assuming that terrorists could gain access to the container). Thus, a demonstration using explosive to blow the lid off and spread a readily identifiable dye far and wide would serve to show just how much harm would result were the dye radioactive.

Either of the imagined demonstrations described above would lead to a conclusion exactly opposite to that advanced by Sir Walter Marshall. They would both suggest that it is dangerous to transport spent nuclear fuel by train. Yet neither of the imagined demonstrations would be *experiments* on the safety of rail transport for nuclear fuel, and they would no more prove decisively that it is unsafe than the CEGB's demonstration proved decisively that it is safe. Nevertheless, by imagining alternative kinds of demonstration, we are better able to see what these demonstration involved; we can see what a demonstration proves and what it does not prove.

The same applies to the plane crash. We can easily imagine the demonstration going according to plan, the Boeing 720 landing wheels-up without bursting into flames, and the TV cameras entering the fuselage of the unburnt aircraft and revealing pictures of putatively smiling unscorched dummies. Yet that scene too hides some hidden questions. Could it be that the particular crash was unusually benign so that AMK has the potential to make a difference on only very few crashes? Could it be that the extra machinery needed to turn jelly-like fuel into liquid before it enters the engines of jet planes would itself be a cause of crashes? Could it be that the transition period, requiring two types of fuel to be available for two types of plane, would be so hazardous as to cost more lives than the new fuel would ever save? Could it be that the extra cost of re-equipping airlines, airports, and airplanes, might jeopardise safety in other ways at a time of ruthless competition in the airline market? Again, perhaps there was too little time for anyone to escape between first and second fires; perhaps cabin temperatures rose so much during the first fireball that everyone would have been killed anyway; perhaps there is always unburnt fuel left after a major crash; perhaps everyone would have died of fright. Both the analysis and the conclusions to these questions need more expert input than we can provide.

CONCLUSION

We started this chapter asking whether the solution to experimental ambivalence was to be found in great public spectacles. In the case of the experiments we described in *The Golem*, and in the case of the other technologies described here, experts, given the most detailed access to the events, disagree with one another, and an outcome to a debate may take many years or even decades to come about. Yet in the case of these two crashes it seemed that the general public, given only fleeting glances of the experiments, were able to reach firm conclusions instantly. The reason was that they were not given access to the full range of interpretations available to different groups of experts. The public were not served well, not because they necessarily drew false conclusions, but because they did not have access to evidence needed to draw conclusions with the proper degree of provisionality. There is no short cut through the contested terrain which the golem must negotiate.

NOTES

[1] Greenpeace claim that their 'leak' as regards the CEGB's earlier programme of experimental work came from inside the CEGB itself, and from an independent engineering source close to Ove Arup, the firm that ran the test for the CEGB. They say that when their claims were presented to an authoritative seminar organised by the Institute of Mechanical Engineers, their arguments were not proved unfounded except in one small respect. (At the time, Greenpeace claimed that the bolts securing the engine of the locomotive had been loosened by British Rail. They retracted this later.) I have not done any 'detective work' to test Greenpeace's claims, so I present this information merely to suggest further that the counter-interpretations put forward by Greenpeace were technically informed and can be held by reasonable persons.

[2] The programme was called *The Real World*. A remarkable edition of this entitled 'Up In Flames' was broadcast on Independent Television.

'Up in Flames' is an interesting TV programme in the way it re-interprets scientific evidence. The handling of scientific evidence and doubt is very sophisticated yet still fascinating. Nevertheless, the programme draws back from forming a general judgement about the widespread flexibility of the conclusions that can be drawn from experiments and tests. The programme presses the FAA view on the viewer. We do not endorse the FAA view but use the programme as a means of revealing the alternative interpretations than can be provided by different sets of experts.

[3] Private communication with the author. Interestingly, the ICI spokesman used the terms 'demonstration' and 'experiment' without bidding.

[4] Private communication with the author.

4

The world according to Gold: disputes about the origins of oil

Every schoolchild sooner or later learns the standard story of the origins of oil; it runs something like this. Once upon a time, hundreds of millions of years ago, the earth was covered by vast oceans. Animals, plants and micro-organisms in the seas lived and died by the billion, their remains sinking to the bottom and mixing with sand and mud to form marine sediment. As the ages passed, the mud turned to rock and eventually the organic mass became buried deep under layers of rock. The oceans receded and the earth's crust heaved and buckled. Compressed under this vast weight of rock, decomposition occurred and the layers of biomass underwent a chemical change to form hydrocarbons (compounds composed only of hydrogen and carbon atoms) – coal, oil, and natural gas.

Special geological conditions are needed to keep the oil trapped underground. The organic material has to be covered by porous rocks and these, in turn, have to be covered by an impermeable layer which acts as a cap to prevent the oil and gas escaping. Oil is consequently found only in places where these geological conditions are met.

Although this crude, even mythical, account has become greatly more refined in modern petroleum geology, the underlying tenet that oil is formed by biological decay is the starting point for any exploration of the subject. And we all know that there are good reasons why oil is called a *fossil fuel*: the large numbers of fossils found in and alongside coal, and the micro-organisms found in oil are taken to demonstrate their organic origins.

Perhaps the most compelling reason to believe that we have it right as regards the origins of oil is the commercial interests involved. There is a giant petro-chemical industry which derives its profits from mining and refining 'black gold'. Oil is at the roots of our modern industrial civilization. It is not some esoteric sub-atomic particle upon which a few physicists have built their careers. When we talk about oil we are talking about no less than the history of the modern world. A vast industry supported by national governments makes sure it understands *how* oil is found, *where* it is found and *who* has the rights to find it. Surely, if all this effort has made us so rich, we must be smart. Or are we?

There is one man on the planet who seriously thinks we have it wrong. His name is Thomas Gold.

GOLD'S WORLD

The world according to Gold is actually not that different from the world we know. Oil would still be found in all the places where it has already been discovered, it is just that there would be more of it, and it would be located in more places. According to Gold, we are wrong about how oil is formed. Oil is not the result of biological decomposition, but is produced non-biologically, or 'abiogenically'. When the earth was first formed, primordial hydrocarbons were trapped deep under the surface. These primordial hydrocarbons, by a continuous process of 'outgassing', have gradually and continuously migrated to the surface. The oil discovered so far is only that which is most accessible. There are, according to Gold, other reservoirs, of vast extent, waiting to be discovered deep below the surface. Even more importantly, in the world according to Gold, oil should be found in places where, according to the standard theory, there should be none. If Gold is right, countries apparently bereft of oil could be sitting upon huge reserves. The geo-political implications would be enormous.

The non-biological theory of the origins of oil has a long history. In the late 1870s, the Russian chemist, Dmitri Mendeleev, best known as the inventor of the periodic table of elements, rejected the, then current, biological theory of the origins of oil and proposed an abiogenic theory. Mendeleev's ideas received wide currency at the time, but as twentieth century geology developed his theory started to lose credence. Since the 1940s nearly all geologists have accepted the biogenic theory. The very tiny group of dissenters has Gold as its most vocal supporter, but also includes some Chinese geologists, some Swedes, Gold's collaborator Stephen Soter, and the oil prospector Robert Hefner III.

WHO IS GOLD?

If Thomas Gold did not exist he would have to be invented. Throughout his distinguished career he has made a habit of entering new fields, ruffling old feathers, and on the way he has laid a surprisingly large number of scientific golden eggs. Gold was born in Austria before moving to Switzerland where he studied physics. He then emigrated to Britain to continue his education at Cambridge University. He worked on radar during the Second World War, then moved to the US where he has spent almost his entire active scientific career. He is a Fellow of the Royal Society and a member of the US National Academy of Sciences.

It was Gold who first came up with the term 'magnetosphere' and did much of the early research on it. More famously, he was part of the team that produced the steady state theory of cosmology (Fred Hoyle being the better known part of the partnership), that for years was taken as a serious rival to the 'big bang'. Long before biophysics became a popular subject, he treated the human ear as an active receiver. Most stunning of all, he was the scientist who first proposed that pulsars were collapsed neutron stars. Gold claims his colleagues denied him speaking time at a conference on pulsars when he first came up with the idea; within a year the rest of the scientific community were proved wrong and Gold's paper on pulsars has become a classic. In between these interventions into other fields, Gold does some, by his own standards, quite ordinary things. Indeed, Gold himself contends that the vast majority of his work is mainstream, widely accepted and widely shared; only in these high-publicity incidents has he been on the fringe. Gold accepted a professorship at Cornell University in 1959 and went on to oversee the construction of the world's largest radio telescope at Arecibo, Puerto Rico. Today Gold is as active as ever and holds an emeritus professorship in Space Sciences at Cornell University.

There is no doubt that Gold is widely respected for his scientific abilities. As one scientist, who is otherwise completely scathing about Gold's work in geology, remarked:

> Thomas Gold is, by the way, a very imaginative and wonderful scientist.
>
> *(Cole, 1996. p. 736)*

Another commented:

> Tommy has had many extraordinary ideas that are not in the mainstream, and very often they've proven to be correct . . . He's just very, very, good.
>
> *(p. 737)*

Gold is known for his debating prowess and charisma:

> In fact, no one from this side can successfully debate Gold. I shouldn't put it that way. I think people debate Tommy, but Tommy is really a very persuasive person. It's just his arguments are not persuasive to people from the earth science discipline in general.
>
> *(p. 738)*

Or as another person sceptical of Gold's abiogenic theory declared:

> I wish I could put across what I believe is right as well as Tommy Gold can put across what I believe is wrong!
>
> *(p. 738)*

WHICH CAME FIRST, THE FOSSIL OR THE FUEL?

For any non-biological theory to be taken seriously it must be able to explain why organic molecules and other evidence of organic life are found in oil. It is here that Gold, rather than being put on the defensive, can claim support for his ideas.

Geomicrobiology, the study of subterranean bacterial life, is one of the newest and most exciting fields of science. Microbes have been found living in the most unlikely of places, like geysers, deep ocean vents and even oil deposits. They live at improbably high temperatures (up to 115 °C), improbably great depths (down to 5 kilometres), and they feed off improbably poisonous compounds such as sulphur and arsenic. There is, then, life a thousand leagues under the sea.

This life has left traces, traces which Gold uses to bolster his abiogenic theory. The 'biological marker molecules' or 'hopanoids' found in petroleum are usually taken to be one of the central planks supporting the biogenic theory. How could these extremely complex organic molecules have got into petroleum unless petroleum is itself the product of some sort of organic decay? Gold holds that these molecules, rather than being the remnants of ancient terrestrial life, are the traces of a living subterranean biosphere which uses petroleum as a chemical energy source. For most geologists the fossils feed the fuel, but for Gold, the fuel feeds the fossils.

Gold has recently (1997) linked his theory of biological life deep within the earth's surface to the controversial claims to find traces of microbiological life in a Martian meteorite. According to Gold's theory, micro-organisms metabolize hydrocarbons and they can do it, not only within the earth, but also on other planets. Gold told the 1997 meeting of the American Association for the Advancement of Science, 'Down there, the earth doesn't have any particular advantage over any other planetary body'. He went on to point out that there are at least ten other planetary bodies that have enormous amounts of hydrocarbons in their atmosphere.

As is often the case, this scientific controversy over the abiogenic origins of fossil fuels is not a straight two-party fight between Gold and his critics in geology. Other disciplines are involved. It turns out that the microbiologists, unlike the earth scientists, are somewhat more receptive to Gold's theory. Unlike their colleagues in geology they do not have a direct stake in how oil is formed. As one microbiologist put it:

> He's obviously threatening the general community of petroleum geochemists who have built up a whole discipline on the basis of certain hypotheses which he's trying to tear down. And that can be very threatening. And that's most likely the reason for resistance there. From my point of view as a microbiologist what he's proposing is not at all threatening. It's much more like 'well, let's find out more about that, if we can'.
>
> (Cole, 1996, p. 744)

Gold himself acknowledges the strategic importance of enlisting the support of the microbiological community and thus in making a run into the end zone around the geologists.

... it will sort of outflank them a little because once this view becomes generally accepted, that there is very widespread life below us, then, of course, they will eventually understand that the argument about the biological origin of oil has been greatly undermined by that. Or destroyed. So, yes, it does outflank them, but with a benign audience.

(p. 744)

Gold's strategy here also draws upon the one great revolution that the earth sciences have experienced – and one which still embarrasses many geoscientists – plate tectonics. This is the idea that the continents once formed one big land mass which, over the aeons, slowly drifted apart. The theory was only accepted in geology after being ignored and resisted for decades. Alfred Wegener, the chief proponent of plate tectonics, was often dismissed as a climatologist who had no business doing geology. Wegener enlisted the support of people from peripheral disciplines. For example, palaeontologists who noted similarities between the species found in West Africa and in the eastern part of Latin America were able to provide crucial support to Wegener, thus piquing the geologists:

> Wegener in proposing Drift in effect also proposed a redefinition of geology and a reconstitution of the disciplinary field which included elements of what were otherwise considered heterogeneous bits of disparate disciplines; e.g. geophysics, biogeography, climatology and palaeobotany, and which moved these from the wings to the center stage.
>
> *(p. 744)*

Geologists are sensitive to the comparison between Gold and Wegener, but Gold's evidence has not convinced them and unlike Wegener's has not got better over time. As one petroleum geologist remarked:

> We often feel guilty that we once were wrong, so we may be a second time wrong. So we should be very, very careful about rejecting things which seem to us ridiculous, but the big difference between Wegener and Gold is Wegener had geologically very good evidence for his hypothesis.
>
> *(p. 737)*

Part of the reason geologists feel so confident that they have it right comes from simple ideas to do with the porosity of rock and its permeability to oil. Given that rock deep under the surface is under great weight from the rock above it, it is difficult to understand how the rock could be porous enough or permeable enough to allow fluid flow to, say, fill a well drilled deep underground (as we shall see later, oil found in deep wells forms a crucial part of Gold's case). Although there have been anomalous flows of fluid (water) reported in deep wells it is difficult to understand how these could be sustained over geological time periods.

ABIOGENIC EVIDENCE?

What evidence does Gold have?
Doing science on the origins of oil will always involve drawing large conclusions from small amounts of evidence. Events that occurred millions of years ago must be reconstructed. In some ways it is easier to delve into the first few nanoseconds of the universe than to reconstruct something buried deep in the core of the earth. There is an asymmetry between our ability to look outward from earth through nearly empty space and our ability to look inward. We have no direct equivalent to the telescope for looking through solid matter and hence we must surmise what happened millions of years ago from the comparatively thin surface crust to which we have access and from evidence from volcanoes and earthquakes which reveal tantalizing glimpses of what lies below. Although seismic, gravity, and magnetic data, along with inertial measurements tell us something about the earth's deep interior they do not provide such compelling evidence as Galileo obtained from the heavens by simply pointing his telescope at the moon and planets.

Gold claims that when the earth formed, 4.5 billion years ago, hydrocarbons accumulated as solids. They have since slowly seeped up through the mantle in vast quantities to the surface via 'outgassing'. The presence of hydrocarbons on other planets in the solar system is claimed by Gold as one of his most telling pieces of evidence. Since these other planets have never, as far as we know, hosted an active ecosphere, their hydrocarbons are presumably of a non-biological origin. Why should terrestrial hydrocarbons have been formed in a different way?

Geologists, in response, point out that the hydrocarbons found on other planets are mostly methane. They have known for a very long time that traces of methane are found in volcanoes. So for them there is no mystery about the source of the methane found on other planets. Methane may well be formed deep within the earth, but methane is not crude oil.

Often oil is found in one geographical region despite differences in the geological and topographical terrain and the age of the rock. The simplest explanation, according to Gold, is not oil-producing conditions repeating themselves over time in the same place, but a deep-seated source of the oil. This seepage to the surface could be a cause of earthquakes and explain why historically they are associated with fires, flares and petroleum odours. The upward percolation of oil and gas also explains anomalies in the amounts of trace metals found in oil and the disproportionally high concentration of helium-three found in natural gas – these elements originate from deeper in the core according to Gold's theory.

Again we must stress here that Gold's views are highly unorthodox. One geologist has pointed out that no one on the highly regarded USGS Earthquake Prediction Board supports Gold's ideas on the causes of earthquakes. Also the standard response is that helium-three is a much more mobile gas than methane and hence finding anomalous amounts is not that surprising.

Furthermore, Gold is regarded as someone who lacks experience in geology. The geologists can cite an enormous body of literature and evidence, all supporting the biogenic theory.

A CRUCIAL OIL WELL?

The debate seemed interminable until, in 1985, Gold came up with something which had the potential to settle matters for all time. He proposed a crucial experiment. The most direct way to test the abiogenic theory would be to drill an oil well in a place where there was no sediment and where no oil should be found. If oil was found, then in classic Popperian style, the biological theory of oil would be falsified. Finding oil would surely, if nothing else, be a clean-cut piece of evidence. As Gold remarked, '[It] was just such a clean case'(Cole, 1996, p. 745).

In 1985, at Gold's urging, the Swedish National Power Company, Vattenfall, agreed to begin oil exploration in the Siljan Ring, a giant meteor impact crater in a huge mass of granite in the middle of Sweden. In the summer of 1987 the first well, Gravberg-1, was drilled. Much to everyone's surprise one hundred litres of black oily gunk was pumped out along with some methane. Had Gold, by finding this small amount of oil-like substance in a place where none should be found, refuted the biogenic theory?

Unfortunately for Gold the process of falsifying theories is not as straightforward as some thinkers would have it. As soon as the oily gunk was found, supporters of the biogenic theory came up with a new interpretation of what had gone on. They argued that the oil pumped out of the well was none other than refined oil, probably diesel oil, an ingredient of the drilling mud used to lubricate the drill bit! The oil that was found had, according to them, actually been pumped down the well four months earlier when drilling began.

One might think that this disagreement over where the oil came from could have been resolved using analytic techniques such as gas chromatography and mass spectroscopy. Also surely anyone with a match and a sense of smell could tell the difference between fetid, often highly-flammable crude oil and refined diesel which does not easily ignite? But, as is often the case with golem science, things turned out to be much more complicated. When the analysis was completed it indicated the presence of certain chemical markers consistent with gilsonite, a drilling mud additive (there are other petrochemical components in drilling mud). This led some geochemists to conclude that no new oil had been found. But when the gas chromatography was completed, the recovered oil not only did not match that of refined diesel but, in addition, trace metals were found in higher concentrations than occur in diesel oil. What of the flame test? According to at least one report, the recovered oil was 'more easily ignitable than the diesel, and it had a different odor' (Cole, 1996, p. 746). Unfortunately no one anticipated this dispute and no sample of the drilling mud had been taken before it was poured down the hole. The geochemists were forced to turn some interpretative somersaults in order to maintain their view that the oil came from drilling mud. They talked about possible chemical alterations to the diesel downhole, such as the enrichment of gilsonite, or even possible microbial activity which changed its constituents!

The data were ambiguous. The preferred conclusion depended upon which scenario seemed the more plausible: that there is naturally occurring oil in granite formations, or that drilling fluid undergoes some sort of chemical or biological metamorphosis downhole.

Gold, himself, was dissatisfied with this first experiment. Indeed he refers to it as a 'complete fiasco'. He redesigned the experiment substituting water-based drilling mud for oil-based drilling mud. A second well was sunk, Stenberg-1. Again methane and small, but significant, amounts of oil (12 tons of oily gunk or about 80 barrels of oil) were found. Gold hoped now, that if he had done nothing else, he had at least refuted the biogenic theory.

Immediately, however, supporters of the biological theory came up with a new explanation for the oil. The thin layer of sediment that covers the Siljan granite was claimed to be the source for the hydrocarbons. The suggestion was that the hydrocarbons had actually originated near the surface and had seeped downwards to form subterranean petroleum reservoirs:

> Paul Philp, a University of Oklahoma geochemist who analyzed examples of the 'black gunk' that Dala Djupgas extracted from the hole before pumping out the 80 barrels of oil, says that he could not distinguish between the samples from the hole and oil seeps found in shales near the surface in the Siljan area. The obvious explanation, says Philp, was that oil had simply migrated down to the granite from sedimentary rocks near the surface.
>
> *(Cole, 1996, p. 748)*

Cole goes on to remark:

> Gold, however, thinks that Philp has his facts upside-down, that the equally 'obvious explanation' for the similarity of samples is that oil and gas are seeping *up* to the surface.
>
> *(p. 748)*

Gold told Cole when asked about this issue:

> They would have it that the oil and gas we found down there was from the five feet of sediments on the top – had seeped all the way

down six kilometers into the granite. I mean, such complete absurdity: you can imagine sitting there with five feet of soil and six kilometers underneath of dense granitic rock, and that methane produced up there has crawled all the way down in preference to water. Absolute nonsense! *(p. 748)*

Evidence was also found for biological activity, particularly in an oily magnetite putty that issued from the wells. Gold interpreted this sludge as indicating the presence of deep-dwelling bacteria that subsist on abiogenic petroleum and reduce iron into magnetite.

For the critics, the biological material had rather a different interpretation. It was further evidence that the well had been contaminated from the surface. Within their paradigm it was impossible for micro-organisms to survive at such a depth, therefore the only way they could have got there was from surface contamination, thereby proving that the oil must have seeped down from the surface. Whether contamination did or did not happen depended upon to which theories of oil and micro-organisms you subscribed. As one microbiologist commented:

> It may sound like double talk, but if the contamination problem is controlled and if you then get a sample up from deep underground under high temperature and high pressure and you find the organisms in that sample, then that's very good evidence that they exist deep underground. If the contamination problem is not controlled and you get the results that you're looking for – i.e. bacteria that grow under high temperature and high pressure, perhaps anaerobic bacteria, it doesn't necessarily mean that they're coming from deep underground. They could have come in with the fluids that you were pumping down into the borehole in order to obtain the sample and could result in the contamination. And there's always this little element of insecurity that we have, so that the final way that you use to check, if you can, is: does it make sense for these organisms that you isolate, knowing all the organisms that you possibly could isolate, does it make sense for *them* to be in these samples? And that's where you have to know a lot about the environment that you're sampling.
>
> *(Cole, 1996, p. 748–9)*

This is the 'experimenter's regress'. If you believe that microbiological activity exists at great depths then this is evidence that a competently performed experiment has been carried out. If you believe that microbiological activity is impossible or extremely unlikely then the evidence of biological activity is evidence for doubting the experiment. Experiment alone cannot settle the matter.

GIVE US A GUSHER

Gold's view of the crucial test is straightforward. He believes that he has accomplished what he set out to do – the biogenic theory has been shown to be false. From the point of view of the geologists, however, the matter is also settled; according to them Gold was given a fair chance but he failed to prove his case.

But perhaps even more telling for the critics is the commercial argument which underscores the scientific argument. In the end none of the niceties matter because Gold was not able to deliver a gusher. That, in the last analysis, would be irrefutable proof that his science was correct. And here Gold faces a dilemma. Drilling oil wells is a very expensive means of experimentation. In order to enlist the support of oil-barren Sweden, with its nuclear energy programme curtailed by environmental protests, Gold had to offer the expectation that they would find some commercially exploitable amounts of oil. As he told Cole:

> The Siljan structure is large by any standards. In areal extent, it is like Kuwait. So, if people ask me, 'how much are you expecting to find there', I answer, 'well, you could well imagine that you'll find another Kuwait'. So that was the reason for selecting it, and also, of course, it was easier to go to a wealthy country than to go to a poor country . . .
>
> *(Cole, 1996, p. 750)*

But no investor, however wealthy, wants to sink a well which only brings up oil of great scientific import, they also want commercial success and this Gold has so far failed to deliver.

What is interesting is that geologists sometimes support this subtle raising of the standards of proof by conceding the existence of some abiogenic hydrocarbons, but *not* commercial quantities. As William Travers, Cornell's resident petroleum geologist, puts it:

There is methane that comes out of volcanoes, so you're not surprised that somebody discovers a bit of methane in a crystalline rock. It could have been primary methane and been there in the first place. But it's something else to say this can be produced in very large quantities, enough to supply Sweden with its energy requirements. A lot of gas would have to be extracted from those rocks. And I would say for Gold's theory to be valid one would have to find methane, at a bare-bones minimum – never mind oil – in huge quantities. Otherwise his theory's no good to us. It's nothing we didn't already know. What we don't believe, what we believe is not true, is that it can be found *in huge quantities* and extracted from those rocks *in huge quantities*. So that's the way the theory falls down. The trace amounts that have been discovered so far don't prove anything. They don't prove his theory, and they certainly don't relieve Sweden of any of its energy concerns.

(p. 751)

By reminding everyone of Gold's promise that he could produce a gusher the critics have yet another way of dismissing the results of Stenberg-1. Gold, however, has a nice way of dealing with this criticism. He refers to the story of the chambermaid's baby. When the son of the household was forced to confess to being the father of the love child, he said: 'But it's such a *little* baby'! (Cole, unpublished). Also, if Gold had found a gusher we can speculate as to how the geological community might have responded to it. No doubt they would soon have come up with a convincing explanation as to why *this particular* gusher originated from biogenically produced oil.

Indeed, for one geologist it is the lack of commercial interest in what Gold had found in Sweden which is the most damning evidence. This geologist has pointed out that a 'show' (the term used by geologists to refer to the first signs of oil in a well) of 80 barrels is actually a very promising find for a wildcat. Usually such 'shows' would produce a great amount of interest and the rights to all available land nearby would quickly be bought up for more explorations. That this did not happen in the Siljan Ring, he felt raised questions about the validity of Gold's claims to have found some oil. That commercial standards of proof can run in a very different direction to scientific standards is indicated here. While, scientifically, 80 barrels of oil is no doubt impressive, from this viewpoint

Gold's mistake is to have found *too much* oil – the logic implies that if he had found less then the lack of immediate commercial interest would have made sense. That he found oil without producing a subsequent commercial interest implies something suspicious about this find. Indeed this selfsame geologist pointed to the tradition of pranksters in the oil industry – pranksters who might have turned their humour on Ivy-league professor Gold and his abiogenic theory by pouring the 80 barrels into the well themselves!

In any case, geologists, by playing up the connection between their theories and the commercial exploitation of oil, run the risk of drawing attention to the ways prospectors have actually found oil. Many oil prospectors describe their success as stemming more from art than from science. Indeed, parts of the world's oil supply have been found with scant regard for the prevailing geological wisdom. Big oil fields like Saudia Arabia, East Texas and West Texas were not found using standard geological methods. Oil prospectors cite instinct, as often as science, as the origin of their success. Frank Holmes, who promoted the first oil concessions in Saudi Arabia, 'in the face of the virtual unanimous verdict of the world's leading oil geologists that Arabia would be "oil dry", said his nose was his geologist' (Cole, 1996, p. 753). Wildcatters are known for their disregard of scientific and industry expertise. They recall John Archbold, the chief executive of Standard Oil, who in 1885 rashly promised to drink every gallon of oil produced west of the Mississippi and predicted the imminent exhaustion of the petroleum supply.

Gold claims that petroleum geologists often find oil fields by following fault lines from one producing well to another. Gold says that what prospectors are in fact doing is inadvertently tracing out petroleum generating structures located far deeper than the supposed source rocks:

> One can mention many examples where long lines flow from one geologic terrain into another and continue to be oil rich. And the oil prospecting people know that. They don't know why. But they mostly say 'well, we don't care why. It's of no interest to us why. We just know this a way of finding oil, and we just do it'.
> *(Cole, 1996, p. 754)*

Again, petroleum geologists dispute Gold here, citing independent scientific procedures as the source of new discoveries of oil fields and criticizing Gold for introducing what they regard to be meaningless expressions like 'flow lines'. They think Gold has it wrong and that modern petroleum geology and the accompanying technology provides a much better way of finding oil than methods used back in the nineteenth century.

Gold has attracted a lot of criticism because Dala Djupas, the company that drilled the Swedish wells, has been bankrupted; Gold has been accused of partaking in a swindle, defrauding investors of money. Although Gold strongly rejects such claims, and most geologists Cole talked to in his recent study of the affair were careful to explain that the charges were only hearsay, there can be little doubt that the financial spin-offs have not helped Gold in his battle with the geologists. Gold is indignant about inferences his colleagues might draw:

> The idea that someone with a scientific career behind him would risk destroying it with a gross swindle of this kind is just so utterly absurd.
>
> *(Cole, 1998, p. 13)*

He also points out that financial self-interest can be cited against the geologists working for the oil industry:

> Do you suppose that the petroleum geologist who has been advising Exxon to drill for hundreds of millions of dollars for maybe thirty years, will go to his bosses at Exxon and say, 'I am sorry, sir, but I have been wrong all those years. We have been finding the petroleum, but if we had searched for it in another way, we would have found ten times as much?' . . . It is very unlikely that they will do that.
>
> *(Cole, unpublished)*

One of the most vituperative debates has concerned Gold's participation in a compendium produced by the United States Geological Survey (USGS) to promote the use of natural gas as a cleaner and potentially cheaper alternative to oil. David Howell, a USGS geologist, on hearing about Gold's theories about the origins of natural gas, invited him to submit an article to be included in the book. Howell soon found himself at the centre of a controversy as three of the eight members of his own editorial board resigned claiming that publishing Gold's piece would damage the credibility of the Survey. The book was eventually published but outraged geologists circulated a letter of protest. In this they revived reports about financial misdealings in Sweden and called on the USGS to withdraw the book from libraries worldwide. Howell was shocked by the outcry:

> The most stupefying aspect to me was that university professors signed on with this letter saying that the book should be retracted. Burned! We were supposed to burn the book!
>
> *(Cole, 1998, p. 13)*

Howell also took with a grain of salt the allegations of wrong-doing – he felt it was a 'bit rich' coming from people so close to the petroleum industry:

> I thought, well what the hell has the petroleum industry been about for the last hundred years. You read the history of petroleum exploration, and it's filled with charlatans and guys with crazy ideas that are going out making big promotion and drilling and occasionally striking it rich. And then all the wells that were drilled that were known to be dry, purely for tax purposes. And I thought, this is part and parcel of the petroleum industry. Why are they having such a reaction?
>
> *(Cole, unpublished)*

Howell's own explanation for his colleagues' reaction is that Gold could dent the credibility of an industry bent on spending billions of dollars a year on oil exploration. Whether Howell is right or not, it is clear that issues of what counts as commercial credibility have become caught up with the more usual issues of scientific credibility. In short, Gold's non-biological theory and its assessment are intertwined with the politics and commerce of oil exploration. There is no neutral place where a 'pure' assessment of the validity of his claims can be made.

The remarkable thing about this case is that big, robust, expensive technological objects like oil wells have become subject to the niceties of technical disputes with which we are more familiar from research frontier science. The experimenter's regress applies to oil wells – this, of course, will be no surprise for petroleum geologists because drilling an oil well is to all intents and purposes just like running an experiment. This case, like the earlier chapter on the

Patriot missile, shows the other side of applied science and technology. The uncertainties and interpretative judgements which make up the craft of technology have seeped to the surface.

Whether Gold is right or not we do not know. Certainly most scientists think he is wrong. What is clear, however, is that deciding in a case like this, even when the odds are stacked so formidably against the scientific maverick, is rarely straightforward.

Lastly, in this debate we have seen that there is no neutral terrain where pure assessments of either side's case can be made. For the critics of Gold the experimenter's regress has been closed by appeal to commercial considerations – the fact that Gold has not been able to produce a gusher. But Gold in turn has questioned the actual part played by the standard biogenic theory in the practical world where oil prospectors operate. The burden of proof has been passed back and forth between the worlds of science and of technology. The residual uncertainties of golem science and technology are kept at bay by this ability to pass seamlessly between science and technology.

5

Tidings of comfort and joy: Seven Wise Men and the science of economics

Shortly after the Second World War, an engineer from New Zealand, 'Bill' Phillips, working at the London School of Economics, built a model of the economy. The marvellous thing about this model was that it ran on water. Phillips's model was a set of tanks, valves, pumps, pipes, baffles and cisterns. If, say, the flow into some cistern increased while the cross section of the output remained the same, the water in the cistern would rise. The new level might increase the flow of water into another cistern, raising its level, or it might be enough to trigger a valve and restrict the flow somewhere else. The whole thing, which stood about seven feet high, weighed a good part of a ton, and was prone to leakage and corrosion, was meant to represent the flows of income around a national economy. Changes of levels were linked by indicators to scales which represented measures of economic performance such as price indices, stocks of money, or Gross National Product. It was even possible to link one of these gurgling monsters to another, thus representing the interaction of two national economies, or the interaction of one economy with the rest of the world. Phillips's hydraulic model of the economy has been restored recently and can be seen at the Science Museum in London.

Nowadays no one would dream of building a model of the economy that ran on water. Nowadays one would use a computer and the relationships would be represented by interacting mathematical equations. Using a computer and equations one can build the equivalent of many more pipes, tanks, and valves than one could ever construct with plumbing. This is what macroeconomic modellers do;

they use equations to build a model of the economy. They model not only theoretically derived relationships but quantities based on observations of how this or that change has appeared to affect the economy in the past. Modern models may have hundreds of equations and variables arranged in a big tree-like structure representing everything from world interest rates to levels of business and consumer confidence; the output of some equations will count as variables in other equations, while these effect still other equations and so forth. A modern model rendered hydraulically in the style of Phillips would be big enough to flood the LSE and the surrounding streets.

Phillips's model can itself be thought of as a partial rendering of the mathematical ideas of the famous economist John Maynard Keynes, who did his pioneering work in the 1930s. By all accounts it was a successful rendering, both dramatically and technically. People could watch the water flowing and visualise the ideas of a great abstract thinker. But what counts as success in a modern model of the economy? It is one thing to take a set of abstract ideas about economic interactions and make them easier to comprehend, it is another thing to predict what is going to happen in the British economy over the next few years. The criteria of success are very different; success of one kind is often taken to promise success in the other while in practice there may be no connection at all.

WHAT IS A MACROECONOMIC MODEL MADE OF?

Each equation in a macroeconomic model relates one variable to another. For example, an equation may tell you that consumers' expenditure is related to increases in consumers' income in a certain way. If things were very simple it might be that if income trebles, consumption will double, the remainder being invested. In practice, how much of the extra income is consumed will depend on lots of other factors. For example, it might depend on how the extra is distributed between rich people and poor people. If you are starving and your income trebles you are likely to spend all the extra immediately on food. If you are very rich and your income trebles you may hardly notice it in terms of a change in your pattern of consumption.

Thus, if extra income is distributed disproportionately to the low paid, it is likely to result in more extra consumption than if it goes to the rich. Changes in national consumption will also depend on aggregate levels of confidence about the future: are people generally inclined to 'spend, spend, spend' or to 'put something away for a rainy day'? All these relationships can be expressed in equations which depend on other equations, which in turn depend on other equations, and so forth.

To take the 'spend, spend, spend' versus 'save for a rainy day' example, if consumer consumption, C were thought to be two-thirds of income, Y, then the relationship could be expressed in the equation, '$C = 0.67Y$'. This is but a 'toy' example: here is a real equation covering the same variable taken from a simple modern macroeconomic model:

$$\log C_t = 0.628 \log(C_{t-1}) + 0.315 \log(C_{t-2}) + 0.165 \log(RPDI_{t-1})$$
$$- 0.0879 \log(RPDI_{t-2}) - 0.00365\, RLB_{t-1} + 0.00161\, RLB_{t-2}$$
$$- 0.225 [\log(PC_{t-1}/PC_{t-2}) + 0.0057] - 0.0111 [\log(PC_{t-2}/PC_{t-3}) + 0.0057] - 0.196$$

Key: C = consumers' expenditure; t = time; RPDI = real personal disposable income; RLB is a measure of interest rates; PC is a price index.

The equation shows how consumption in period 't' is related to a host of other things in earlier periods.

Such equations are important because although the general relationships may be matters of macroeconomic theory – for example, one may safely theorise that spending will go up when income increases – the numbers in the equations have to be obtained by looking at how things happened in the past. One has to get these numbers from existing data, but data may be in short supply, or it may be inaccurate. Furthermore, the equations may come out differently if different sets of data are used. Robert Evans reports that, while he was waiting to interview them, one team discovered that, fortuitously, a problematic equation in their model came to 'look better' when the time period over which they took data was extended from 1973 to 1970! Furthermore, research coming from Manchester University suggests that not only is the time period over which data is taken critical, but the 'vintage' of the data is also important. That is, since our quantitative information about what happened to the economy in any specific time-period in the past is continually being refined, data of a different vintage can suggest a very different equation.

Economists also have to make judgements about which equations to include in their model, what their form should be, and how they are to be related to each other. With several hundred equations to play with, this is an area where 'theory' and 'guesswork' are not as far apart as conventional ideas about science would encourage us to think.

Information is also needed about the environment in which the model is meant to work. That is to say, the model will have to use information about the state of the economy that is not generated by the model itself; these are the so-called 'exogenous variables'. To give a simple example, a model of the British economy will need a figure for the population and another for population growth. Models representing the past can use existing data, though they are not always reliable (for example, in the case of population, British censuses have been wrong by up to ten per cent). When forecasts are to be made, the future values of the exogenous variables have to be estimated *before* the modelling can begin. It is easy to see the problem this presents when one considers that exogenous variables include the state of the world economy, the price of oil, undecided government policies, and so forth, all of which are likely to be far more complex to model than the national economy itself.

A few of these uncertainties can be partially accounted for by statistical techniques. One may say of the outputs of certain equations, not that they have definite values, but that they have probable values within a certain range. Thus, instead of saying that if income is 'x', consumption will be '$0.87568x$', it is more correct to say that one is 95 per cent certain that consumption will be between $C - n$ million and $C + n$ million. There are well-known statistical techniques for calculating 'standard errors' based on knowledge of how the data have behaved in the past. From this one can produce corresponding 'error bars' (so-called because, on a graph, points are extended by lines – or bars – which show the range within which the points are expected to lie). In some cases one can assemble all the error bars belonging to all the equations and try to estimate one big error bar for the model as a whole but, as we will see, even this technique underestimates the true error because it takes into account only those mistakes which have a well understood statistical pattern. Unfortunately, even this underestimate gives a result too large to be useful in terms of government policy.

THE SEVEN WISE MEN AND THEIR IDEAS

Politicians, business leaders and investors are interested in movements of the major indicators of the economy of the order of one percentage point or less. A 3 per cent level of growth is significantly different from a 2 per cent level of growth as far as economic policy and investment decisions are concerned. Inflation is an equally sensitive measure. Governments need to predict the future of the economy if they are to understand the potential impact of their economic policies and develop them accordingly. Governments want to know what will happen if they increase interest rates by, say, 0.5 per cent. Will this fatally damage consumer confidence and send the economy into depression or will it be just enough to damp down a worrying inflationary trend? The British Conservative government's flirtation with the European Exchange Rate Mechanism (ERM), which ended disastrously in 1992, was based on estimates that were wrong by this kind of order.

It would be nice for governments if there were a scientific method for making economic predictions. To build a model which 'retrodicts' the history of sets of economic indicators to within a fraction of a percentage point is not difficult, and to this extent most macromodels are equally good. Predicting the future is a different matter. To put the matter pithily: nearly all the macroeconometric models we have in Britain are successful at predicting the past. They are all *similar* when it comes to predicting the future in that they all predict changes in growth rate of between about 0 per cent and 3 per cent; nearly all of them are *different* in that within these bounds they disagree wildly; finally, nearly all models are again *similar* in that they are nearly all wrong nearly all the time – they may predict changes of the wrong magnitude or even the wrong direction. Worse still, the model that predicts the right outcome is nearly always an 'outlier', so you cannot expect to get it right by taking the average of all the models.

Figure 5.1, taken from page 31 of Burrell and Hall's paper in the journal *Economic Outlook*, shows the performance of a variety of models from 1984 to 1992 compared to the actual performance of the economy – the isolated line made up of straight segments. As can be seen, the models failed to predict the large fluctuations in output and no model ever predicted the contraction in output (the section of the bold line below the x-axis), which is found after 1990.

Irrespective of the performance of econometric models (or perhaps to share the blame for future failures) the British Conservative government of 1992 decided to appoint a panel of scientific economic soothsayers. 'The Seven Wise Men', as they became known, were to predict the short and medium term future of the British economy. They were to meet three times per year and debate their efforts, offering a report to the then Chancellor of the Exchequer, Norman Lamont. (The panel was disbanded by Chancellor Gordon Brown in May 1997.)

The original Seven Wise Men were: Andrew Britton, then Director of the National Institute for Economic and Social Research; David Currie, from the London Business School; Tim Congdon, who worked for a prestigious City of London firm; Gavyn Davies, who was Chief Economist for the fabulously successful Goldman Sachs International; Wynne Godley, the maverick Professor of Economics at Cambridge; Patrick Minford, another idiosyncratic academic economist then at Liverpool University; and Andrew Sentance, Chief Economist from the Confederation of British Industry (CBI). (The economists quoted below comprise some of these seven plus Paul Ormerod, from the Henley Centre for forecasting, and Ken Wallis, Director of the ESRC Macroeconomic Modelling Bureau at the University of Warwick.)

These men were quite open about the likelihood that they would disagree with one another and their reports reflect disagreements in doctrine as well as outcome. For example, in the first report, of February 1993, there are four opinions expressed about how the economy works. Minford is a deep believer in the free market. He is very pessimistic about the short term, predicting only 0.2 per cent

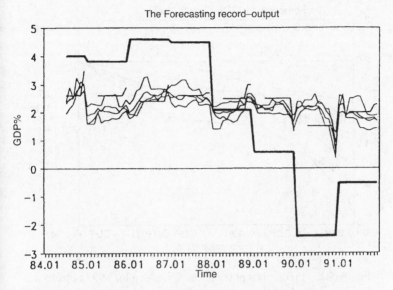

Figure 5.1.   Graphs of average performance of models versus actual economy.

growth for the current year but puts forward an optimistic projection for the medium term, with 3 per cent growth for 1994 and higher figures for several years thereafter. This optimistic scenario depends, according to Minford, on the government directing all their interventions toward safeguarding the self-regulating qualities of markets, including the labour market. Congdon is concerned primarily with monetary policy. He is similarly optimistic in the medium term provided that 'sensible' monetary policies are pursued. Godley takes a radically different position, being primarily concerned with the importance of the balance of payments. He is pessimistic for the short, medium and long term unless measures can be taken that will increase Britain's exports relative to imports (Congdon does not see balance of payments deficits as a problem). The other four economists believe that the economy is producing below capacity, but consider that both the Minford/Congdon optimistic scenario and the Godley pessimistic outlook are too extreme in their different ways.

More detailed figures are to be obtained from tables at the back of the reports submitted by the panel. For example, in February 1993,

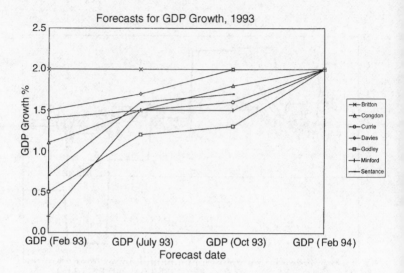

Figure 5.2. Predictions of the Seven Wise Men for 1993 outturn.

the panel tried to forecast the growth and inflation expected during that year. The range of predictions of growth ran from Minford's pessimistic 0.2 per cent to Britton's optimistic 2.0 per cent with figures of 0.5, 0.7, 1.1, 1.4, and 1.5 per cent lying between; the actual outturn was 2.0 per cent. Inflation predictions ran from Sentance's 3.1 per cent to Godley's 4.8 per cent which bracketed the other five's 3.5, 3.6, 3.8, 4.0 and 4.6 per cent; the actual outturn was 2.7 per cent. Thus, the best predictor of growth was Britton. Unfortunately for the science of economic modelling, Britton was also the second worst predictor of inflation and no one was very close in terms of the sort of information that governments need.

The performance of the panel is very nicely set out in figures in the February 1994 report. For example, predictions for growth in gross domestic product diverged wildly at the beginning of the year but, as would be expected, converged on the correct figure as the year progressed (see Figure 5.2).

In February 1994, the economists (now shrunk to six with the loss of Andrew Sentance) were much closer together in their predictions for growth in the current year. The predictions were 2.4, 2.5, 2.7, 2.7, 3.0 and 3.0 per cent. The predictions for inflation were still quite varied, however, being 1.7, 2.4, 3.1, 3.1, 3.2 and 3.8 per cent. The actual outturns were 3.9 and 2.3 per cent respectively. Thus, once more the economy was getting more growth for less inflation than the economists predicted.

WHY DO MACROECONOMETRIC MODELS SURVIVE?

Even if the history of macroeconomic modelling is not encouraging, could the forecasts not be improved? Could not the best features of the best replace the worst features of the worst? The trouble is that before modellers can learn from each other they have to decide which are the best models and which are the worst. Not only is there no agreement about which models are the best, there is no agreement about what it would be to be the best, and no agreement about what it would be for an economic model to fail. How this works out in practice can be seen as we run through some ways in which comparisons might be made. The argument will be illustrated with quotations from economists including some who joined the panel at a later date.

(i) Underlying structure versus the behaviour of the economy
One may believe that economics is a heavily theoretical science like physics. In that case, the role of the modeller is to specify the underlying model; the modeller should not be too concerned about the details because the details have to do with error terms. To use an analogy, macroeconomic models are like Newton's equations. If economic entities float about in space like planets, then the models would be as accurate in predicting their future as Newton's Laws are in predicting the future of the planets. As it is, economic entities are more like, say, the seas; the seas are subject to Newton's laws but because of the wind and weather no one can predict the exact height of next Tuesday's tide. If one thinks of the economy like this, then failures of prediction do not indicate failures of the underlying model. As Patrick Minford put it:

> My starting point for this is that economics is perfectly obviously not a precise science in the normal sense of the word. It is highly

stochastic [there are many random factors]: relationships are very
hard to pin down precisely . . . Nevertheless, there are certain
fundamental laws, basically the laws of supply and demand, which
regulate economic behaviour, particularly over the long term.

On this interpretation a model should be stable over time even if the
actual economy fails to conform to it.

There are other modellers who believe that the economy is continually changing in significant ways but consider their task is, nevertheless, to predict it. These modellers, and they include Andrew Britton,
believe that a model that is too unchanging must be a bad model.
They believe they should be able to spot potential changes. To use
our solar system analogy, they feel they should be able to predict the
future of the planets even if massive comets continually pass through
the solar system upsetting their orbits. The only intractable problem
is when brand new types of disturbing feature appear. Under this
version of modelling, economists expect to include their own judgements in their models.

These two approaches were summed up by Andrew Britton:

I think there are really two different approaches. One is to say that
this is a branch of science and that everything must be based on
objective criteria which people can understand. The other is to say
that is just too inflexible, and that there's something called
judgement – intuition if you like – which has its place in the sciences
and that it's the people who are intuitive who are successful.

(ii) Quantitative prediction

Once one has redescribed the goal of the economic modeller under
either of the two approaches outlined above, one has a licence to be
wrong. The 'stable theory' approach entitles one to be wrong because the model is meant to represent only the underlying structure
of a varying reality; the 'judgement' approach allows one to be
wrong because there is never any expectation that one is developing
the kind of truth, accuracy, and replicability which we expect under
the standard model of science. Wynne Godley told Robert Evans:

I think tables of numbers are the enemies of good forecasting.
[Forecasts] should be judged by whether or not they give a good idea

of what the whole situation is going to be like – what character it
will have.

Another econometrician said of the published forecasts: 'I think the
evidence in the economics community is that, by and large, they
don't take forecast failures as crucial experiments'.

This kind of comment is justified by the reasoning of Andrew Britton:

> [Forecasts] are all probability statements. All econometrics is based
> on the idea of probability criteria, that you are going to say that, you
> know, the likelihood of this result not being consistent with theory is
> below 5% or something, and you have to cut off at some point. You
> don't get the one definitive experiment which shows you that, you
> know, the speed of light is the same in all directions.[1]

Furthermore, because the models can be very sensitive to small
changes, one can always always find data to explain a specific
forecasting mistake while preserving the status of the model as a
successful probability statement. Minford believes that it is only 'big
forecast errors' (see below) that cause people to lose confidence in
models. He put it this way:

> I think most econometrics is rubbish frankly. There's an awful lot of
> sheer bullshit published in the journals where people purport to have
> tested something. But it's baloney, because there's another five runs
> on the same data with slightly different specifications, with slightly
> different sample periods, which have either supported the [null]
> hypothesis or only marginally discounted it.

Another of Evans's respondents suggested that the data base was too
small to make reliable inferences. He said that while economists use
data stretching back 30 years, a base of 300 years would be more
appropriate.

It is not only that the models are sensitive to changes in the
empirical data base on which the equations are based, they are also
sensitive to slight changes in the definition of terms. Economic variables, even those that are firmly entrenched in the heart of theories,
can be thought of in different ways. Thus, Paul Ormerod:

> Even within the same political framework, what is the relevant
> definition of wealth in a consumption function? There are many

different ways of looking at it. I mean, is it important to consider the impact of inflation on income or whatever?

There are many more *ad hoc* reasons that econometricians can supply to explain their failures.

(iii) Big forecast errors

Patrick Minford, as mentioned above, believed that because econometric tests were too easy, deep structural mistakes within a model can be revealed only by 'big forecast errors'. He said:

> There are two main ways in which the profession changes its view. One is if a theory makes palpable nonsense, they jettison it, or if another theory comes and makes sense they may accept it . . . the other things that prove irresistible are big forecast errors.

Minford's model, however, failed to predict the British economic recession of 1980–81 and Evans asked him why this did not cause him to abandon the model since it would seem, on the face of it, to be a pretty big error. He replied:

> Sure, sure, but I'm not talking about, I'm talking about the big errors, the big errors. Now, of course, we didn't call the recession we saw in 1980 for example. We forecast that the Thatcher policies would deliver a mild recession, a growth recession, but a tremendous drop in inflation, but other people were saying there'd be a something more of a recession but no drop in inflation.

Here, Minford is excusing his model's failure to predict a massive drop in output on the grounds that his overall picture was correct in that he spotted the fall in inflation. Evans suggests that Minford could say this because of the prominence that was given to inflation as a measure of success by the Conservative governments of the 1980s. Clearly, if governments wish to draw attention away from massive drops in output and rises in unemployment, failures to predict them will not be broadcast; these indicators were not the targets of Conservative government policy at the time.

(iv) Luck

It is also possible to argue that modellers who did not suffer from big mistakes were lucky while some others were unlucky to have been wrong. If you believe that you are supposed to be modelling the deep structure of the economic system, then getting the right result for the wrong reason is less worthwhile than getting the wrong result for the right reason. As Minford puts it, talking of Wynne Godley's Cambridge group:

> Cambridge . . . got unemployment right because they were very bad on output of course . . . they were too optimistic on their unemployment–output relationships, so they got unemployment right but they got the mix if you like, rather than unemployment, wrong, like everybody else. So I don't think people were terribly impressed by the Cambridge thing, they just felt that they were just far too pessimistic on demand and output and felt that they got unemployment right by mistake.

Likewise, David Currie said this of Tim Congdon's triumph in forecasting the inflations of 1989 and 1990.

> You need to go back with Tim. I mean Tim forecast the inflation of 88–89, but he also forecast it back earlier. I mean he was forecasting that for some considerable time, [and] if you go look at the growth of broad money you see why, because it was growing very fast for some time without causing any explosion. So the question you have to ask is that a forecast, is that a forecasting triumph, or not? [That is, if you carry on forecasting the same thing for long enough, you are bound to be right eventually.]

Even more interesting is Patrick Minford's suggestion that the government were lucky when his pessimistic forecast of a 0.2 per cent growth in 1993 failed to come about, there being a 2 per cent growth instead. He said in the February 1994 report of the Seven Wise Men, '. . . the government have been fortunate in that a still poor situation in terms of delayed recovery is not worse still.'

(v) Trouble with the economy

We see, then, that it is quite hard for macroeconometric modellers to be wrong. If you choose to be concerned with stable structures rather than detailed movements you will not feel accountable to quantitative errors. In any case, everyone agrees that small errors are to be expected in the light of the probabilistic nature of models, the possibility that input data may be wrong, the fact that certain terms can be defined and measured in different ways, the fact that future exogenous variables will have to be guessed at, and that a degree of judgement is inevitable. Even if you believe that large errors are bound to prove you wrong, you may still argue about the meaning of 'large' and you may still think that the difference between accuracy and inaccuracy was not clever economics but luck. Finally, you may always say that the economy changed radically. As Andrew Britton put it to Robert Evans:

> The fact that virtually all the models, all the sort of formal fully developed models failed to predict, suggests that it was not that our model was particularly bad, but that the underlying economy had changed.

To put this another way, Britton is suggesting that it was not the model but the economy that was wrong.

DISCUSSION

It seems that the macroeconomic modellers suffer from a particularly aggravated form of the experimenter's regress or what we shall call the 'technologist's regress'. The experimenter's regress occurs when scientists cannot decide what the outcome of an experiment should be and therefore cannot use the outcome as a criterion of whether the experiment worked or not. In the case of macroeconomic modelling, at first sight there would seem to be little doubt about what the outcome of a model should be, because the correct outcome turns upon the date of the economist's forecast in the form of a certain growth rate or rate of inflation. But even though we know the outcome, economists still fail to resolve the regress because they cannot agree about what 'agreement' between model prediction and outcome should look like. Firstly, they cannot agree what a match should look like in quantitative terms – should there be broad agreement about trends or narrow agreement about percentage points? This is a problem for all sciences that involve the matching of theoretically derived numbers with experimental results, and it has been brilliantly set out by Thomas Kuhn in his paper on the function of measurement. Secondly, they cannot agree whether numerical matches and mismatches represent more than luck.

In any case, economists know that models are unreliable. This is crystal clear if we turn to their comments on the possibility of setting out the statistically calculated probability of their being correct. Patrick Minford made the following comments to Robert Evans:

> The trouble is that these stochastic simulations, they have a very restricted bunch of errors that they draw on and you've got the, you have the errors in the exogenous variables [variables which are outside the model and are matters of judgement] as well which are not generally included in these stochastic simulation exercises. [Evans: And if they are included the outcome would be even worse?] Oh absolutely, that's why it's absolutely pointless to publish these forecast error bands because they are extremely large. . . . I'm all for publishing full and frank statements but you see the difficulty [with] these standards errors is that they're huge. If you were properly to draw out the uncertainty surrounding a forecast, it's huge, absolutely huge.

Paul Ormerod responded in like vein:

> . . . it's been technically feasible to calculate the mean and standard errors for many years. In fact, we could have done this at the National Institute in the mid 70s, but we suppressed it on the grounds that the standard errors were so large, that it would have been difficult for non-specialists, you know people using the models, using the forecasts, to appreciate. It would have discredited them.

Thus, when David Currie suggested to Evans that it was difficult to assign blame for the failure of his group to predict the consequences of joining the European Exchange Rate Mechanism (ERM), he was being too ambitious. He said:

> We were strong exponents of membership of the ERM, and we thought that it was sustainable. It turned out not to be sustainable. Well that may be because we chose the wrong rate, it could be that fixed exchange rates are a disaster as Patrick Minford would argue,

or it could be because the Bank of England made a lot of tactical errors, and it's easy to argue that. It's very hard to know.

It looks rather as though he should have said that there was not the slightest justification for his thinking that any modelling process could give one any idea at all about whether joining the ERM would be a good thing.

In the light of this analysis we have to ask ourselves what is going on. Let us first try a cynical approach. Let us see things from the point of view of a politician. Politicians are not very good at predicting the future of the economy as the U-turn over the ERM showed. It would be nice if there was a scientific method for doing the job because it would relieve politicians of some of their decision making responsibility. The models, as we have pointed out, have a scientific gloss about them. The equation on page 95 above, is redolent with science. Science is often used as a way of avoiding responsibility; some kinds of fascism can be seen as the substitution of calculation for moral responsibility. Even if the politicians did not themselves believe that the models could deliver a reliable outcome, it would be nice for them if the public did believe it. If the public believed there was science going on, the politicians would be able to share some of the blame for failures with the economic scientists.

Politicians might employ economists for either of these kinds of reasons, depending on their level of sophistication. One suspects that in the early years of Thatcherism there was a serious belief in the power of economic science; the way that Thatcher's early administrations cleaved to the set of academic ideas developed by Milton Friedman, in spite of the massive decreases in production and increases in unemployment, suggests that the doctrine was felt to have a scientific warrant and that the government was driven by a scientistic ideology. Perhaps later Conservative administrations, in which ministers were notoriously unwilling to resign, were more interested in passing the blame to professionals.

Continuing with a jaundiced viewpoint, let us turn to the economists. Academics in Britain and a number of other countries have lost much of the esteem in which they were once held. Salaries have plummeted, job security has been reduced, conditions of service have worsened, and the resources for doing interesting research have been cut to the bone. Even more than in the past, those modellers who were primarily academics might have been ready to accept 'fifteen minutes of fame' as some sort of reward for their efforts. The chance of being a member of a small, high-profile, government advisory team must have been all but irresistible even if it was clear that the strengths and weaknesses of economic forecasting would be exposed as never before. It would be understandable, then, if the economists were not as reluctant as they might have been to have their role as scientific soothsayers to the government so exposed.

Let us now try to put a more positive gloss on the story. Perhaps in expecting econometric models to be accurate we are asking too much. After all, we do not expect the weather forecast to be accurate for more than about a day ahead, yet we expect macroeconomic modellers to make predictions years in advance. Weather forecasting and econometric forecasting are similar in that they try to model extremely complex systems with sets of interacting equations. Why should we expect one to work when the other does not? This still does not answer the question of why politicians should seek the expertise of an impossible science and why economists should continue to offer it. Neglecting self-interest, could it be that econometric modellers are still the best source of advice on the economy in spite of their failures?

Putting the matter in terms of expertise may resolve the enigma. Perhaps the science of macroeconometric modelling is of only indirect value. Think of the sort of tasks given to soldiers in military prisons. For example, in British military prisons soldiers must polish the small pieces of brass on the end of the webbing straps of their khaki knapsacks until they shine like gold, but no polishing compound must be left on the webbing. This difficult task is counterproductive in terms of the military usefulness of the knapsack, because a shiny buckle spoils the camouflage, but in polishing their brasses the soldiers are learning another very difficult skill which is militarily vital – the repression of their personalities. Brass polishing is not meant as a direct analogy for econometric modelling but as an example of how doing one thing may equip one indirectly with a different and more useful set of skills.

We cannot pass without mentioning that it has been said that big financial institutions like to employ those who have taken economics degrees at university less because of their knowledge of economics and more because of other characteristics. Michael Lewis, in his book *Liar's Poker*, puts it as follows:

> Economics, which was becoming an ever more abstruse science producing mathematical treatises with no obvious use, seemed almost designed as a sifting device. The way it was taught did not exactly fire the imagination. I mean, few people would claim they actually *liked* studying economics; there was not a trace of self-indulgence in the act. Studying economics was more a ritual sacrifice. . . . I saw friends steadily drained of life. I often asked otherwise intelligent members of the pre-banking set why they studied economics, and they'd explain that it was the most practical course of study, even while they spent their time drawing funny little graphs. They were right of course . . . It got people jobs. And it did this because it demonstrated that they were among the most fervent believers in the primacy of economic life.
>
> *(p. 26–7)*

Effectively Lewis is saying that a training in economics at a university can be the equivalent of learning to polish your brasses. That is, it selected those who were '. . . willing to subordinate their education to their careers'. It may be, then, that even though the models that result from macroeconomic modelling are useless, the effort of building them equips the modellers with useful skills. It may be that macro-economic modellers, as a result of the prolonged attention to the economy that is required of those who are to have professional credence in the modelling community, know more about the working of economies than most other people. Looked at this way it makes perfect sense for politicians to seek their advice and for economists to offer it with justifiable pride. It is not the scientific conclusions that are the useful end product of modelling, but the wisdom and experience that are the by-product of the rigour enforced by modelling; these people know how one thing in the economy might relate to another.

The message of the Golem series is that the value of expert advice should be judged on the expertise of the advisors, not on the scientific appearance of their procedures. That macroeconomic models are built of trees of interacting equations is actually misleading; it gives the models a credibility that they should not have. It encourages governments to carry on with economic policies long after the actual goings-on in the economy have shown that the advice has been inaccurate – as all expert advice will be inaccurate from time to time. The economists who build the models deserve credibility, but their models do not; one should not use the same criteria to judge expert advice as one uses to judge the coherence of a model.

POSTSCRIPT

Economists and others like to ask, 'if you're so smart, why aren't you rich?' The implication in the jibe works the other way too; many macroeconomic modellers are rich. Does this mean their modelling activities are more effective than this analysis would suggest? Not necessarily.

The way to get rich is to know more than other people about which way the economy is going to go, or know it earlier. That way you can get in ahead of changes in the market. Our theory suggests that the sort of skills that modellers develop do not give rise to forecasts that are reliable enough to allow one to make money. The mistake here, however, may be twofold. Firstly, the way to make money may be a matter of forecasting very short-term changes rather than the medium-term changes that governments want to know about. Econometric models, like models of the weather, may be better in the short term than the long. Perhaps more pertinent, short-term predictions may be a matter of who you know rather than what you know. In other words, proximity to the markets enables one to keep a little ahead of those whose information comes to them second-hand. Perhaps most rich economists work close to the markets.

Secondly, some forecasters are so important that their forecasts change markets. If you are a powerful forecaster, and you know that you are going to predict changes that will have the effect of pushing industrial confidence up, it makes sense to act on this prediction immediately because the forecast is an element in the change that is being predicted; the more people believe you, the more likely is it that your forecast will come true.

NOTE

[1] This mention of the speed of light is apposite. The experiment to which Britton is making implicit reference is the Michelson–Morley experiment. As we show in *The Golem*, the Michelson–Morley experiment was anything but clear-cut in the way that Britton thinks it was. Nevertheless, it is this model of science which informs so much of our thinking and, in the last resort, justifies the use of econometric modelling in spite of Britton's protests.

6

The science of the lambs: Chernobyl and the Cumbrian sheepfarmers

'We may be on the eve of a new age of enlightenment. When a scientist says he doesn't know, perhaps there's hope for the future!' – National Farmers' Union Local Representative during the radioactive sheep crisis.

(Quoted in Wynne, 1996, p. 32)

The accident at the Chernobyl nuclear power plant in the Soviet Union on 26 April 1986 is one of the defining moments of the nuclear age. It is the worst nuclear accident ever: a melt-down of the core of a reactor, followed by an explosion and fire releasing tons of radioactive debris into the atmosphere. The accident not only killed nuclear workers and firemen who fought to save the doomed reactor, but also condemned many others who lived under the path of the fallout to illness and premature death or a life of waiting for a hidden enemy. The weather, no respecter of nation states, carried its deadly passenger far and wide.

Fallout over Britain

At first, Britain seemed likely to escape as its predominant weather pattern comes from the west. Traditional British scepticism about weather forecasts was confirmed, however, when six days after the accident, torrential rain and thunderstorms over mountains and uplands deposited a charge of radioactive material. The Chernobyl cloud had undergone a 4,000 kilometre journey with virtually no precipitation until it reached Britain. Some areas like the Cumbrian fells (better known to tourists as the high valleys of the 'Lake District'), suffered unusually heavy rain – as much as twenty millimetres in twenty four hours.

Special monitoring stations set up for the doomsday scenario of nuclear war provided the first indications of the poisoned cloud. The heavy rain was alarming since precipitation is a major factor affecting the distribution of radioactivity. Radioactive caesium, one of the main constituents of the Chernobyl fallout, is especially affected; one millimetre of rain can deposit as much caesium in an hour or less, as would be deposited in twenty-four hours on a dry day. This meant the Cumbrian fells received as much contamination in one day as drier areas would have received in twenty days of continuous fallout from wind-borne dust.

The government sounds the 'all clear'

Not withstanding the rain, British government spokespersons and scientific experts declared there to be no significant risk. On 6 May, the Minister for the Environment, Kenneth Baker, assured the UK Parliament that 'the effects of the cloud have already been assessed and none represents a risk to health in the United Kingdom' (Wynne, 1989, p. 13). Levels of radioactivity were found to be 'nowhere near the levels at which there is any hazard to health' (ibid.). The cloud was already moving away and a steady decline could be expected in the insignificant, but slightly raised, levels of radioactive contamination. Because the public concern remained high, however, the Department of the Environment began to produce daily bulletins about the radioactivity levels and speculation in the media remained high. Some citizens took matters into their own hands and refused to buy dairy products.

John Dunster, head of the official government advisory National Radiological Protection Board (NRPB) was more circumspect than government spokespersons, forecasting, on 11 May, that the disaster might lead 'to a few tens' of extra cancers in the UK over the next fifty years. He also avowed that 'if the cloud does not come back the whole thing will be over in a week or ten days' (Wynne, 1989, p. 13). Kenneth Baker announced on 13 May that 'the incident may be regarded as over for this country by the end of the week, although its traces will remain' (Wynne, 1989, p. 13). And indeed on 16 May the daily bulletins on radioactivity levels were discontinued.

Radioactive lambs

In Britain responsibility for the scientific assessment of the risks to the public from eating contaminated food lies with the Ministry of Agriculture, Fisheries and Food (MAFF). The same day that Baker pronounced the incident to be 'over', MAFF scientists found that samples of lamb meat from the Cumbrian fells had levels of radioactivity 50 per cent greater than the official government and European Economic Community (EEC) 'action level'. This is the maximum level of radiation permitted before official intervention is required. Despite the warning signs, the official pronouncements continued to claim that contamination was insignificant and decaying further.

On 30 May, MAFF announced new 'higher readings' of radioactive caesium in hill sheep and lambs, but asserted that 'these levels do not warrant any specific action at present'. The hill lambs were still young and the hope was that the high levels of radiation would decrease naturally before they were taken to market.

All previous scientific and government advice was, however, contradicted on 20 June when the Minister of Agriculture, Michael Jopling, announced an immediate ban on the movement and slaughter of sheep in designated parts of Cumbria and North Wales. This ban was to last for three weeks. But even this unexpected development was accompanied by reassurances that there would be minimal effects because radiation would soon fall to acceptable levels. The language was chosen to allay public fear:

> Monitoring results present a satisfactory picture overall and there is no reason for anyone to be concerned about the safety of food in the shops. However, the monitoring of young unfinished lambs not yet ready for market in certain areas of Cumbria and North Wales indicates higher levels of radio-caesium than in the rest of the country . . . these levels will diminish before the animals are marketed . . .
>
> *(Wynne, 1989, p. 14)*

It was clear to all that the radioactive sheep must have eaten contaminated grass. The contaminated material was expected to pass rapidly through the sheep by normal metabolic processes in muscles, excretion into urine, faeces, etc. It was expected that a short ban would be sufficient. This was to assume that the sheep absorbed no new radioactive materials from the grass. This assumption, as we shall see, turned out to be false.

The optimism of 20 June proved to be short lived. Levels of radiation found in sheep continued to increase and on 24 July the ban in Cumbria was extended indefinitely. Similar bans and restrictions were later enacted for parts of North Wales, Scotland and Northern Ireland. At the height of the problem, about one fifth of the sheep population of the UK, or four million sheep, were restricted from sale or slaughter. In 1988, two years after the original accident, about 800 farms and over one million sheep were still under restriction.

These bans had a dramatic effect on the livelihood of upland sheepfarmers. The sale of spring lambs provides the farmers with their only significant yearly income. The timing of sales is crucial. More lambs are produced than pastures can sustain and excess lambs have to be sold before overgrazing occurs. It was even possible that all contaminated lambs would have to be culled, destroying the long-term breeding cycle.

Farmers lost their income at the same time as the normal ecology of their grassland was wrecked. Perhaps worse from the farmers' point of view, they lost their independent decision-making power. These remote, independent farmers, some of the last groups in Britain to experience industrialization, suddenly came under the jurisdiction of a distant scientific bureaucracy.

The farmers' experience of the scientists was not a happy one; they felt betrayed. They were also most unimpressed with the scientists' arrogance and their vacillating pronouncements. Their faith in scientific expertise was undermined. This is a case where the scientists got the relationship with lay people wrong. Its lessons are important for all of us.

THE SCIENCE OF THE LAMBS

What went wrong? The initial scientific assessments of the threat posed by the Chernobyl fallout on Britain in early May 1986 were wrong because scientists underestimated two aspects of rain-borne radiation.

First, rainwater does not flow or gather uniformly, especially in the uneven upland terrain. Rivulets and puddles can lead to large differences in the amount of radioactivity over distances as small as one metre. The amount of radioactivity found in any particular place may not correspond with the level of rainfall in that place. This variability was not understood when the initial estimates of radioactivity were made. The second, more serious, problem arose from the assumption that radio-caesium would not be absorbed by plants after the initially contaminated grass was eaten by the sheep. The chemical behaviour of caesium in soil was based on how it behaved in lowland, clay soils rather than upland acid, peaty soils. In such upland soils, caesium remains chemically mobile and available for root uptake. Under such circumstances radioactive caesium was continually absorbed into the grass and became concentrated in the sheep.

The scientists had been working with a model in which caesium, once deposited, would wash into the soil and then be locked up by chemical absorption, thus contaminating the lambs only once. The scientists gave flawed advice because they failed to take into account the special geological conditions in Cumbria. That the model was wrong became apparent only over the next two years of continuously high contamination levels.

It turns out that there were actually data available dating back to 1964 on the absorption of caesium in upland soils. But these data were interpreted in terms of the direct physical threat which radioactive caesium posed to humans. The threat became correspondingly less the deeper the caesium was buried. What no one contemplated was that the caesium might pose an indirect chemical-biological threat by absorption into roots and then ingestion into sheep.

The flawed advice alone need not have caused the farmers to lose confidence in the scientists. Scientists occasionally get things wrong and sheepfarmers are used to conditions of uncertainty; they know the capricious nature of the world. What particularly dismayed the farmers was the overweening certainty with which the scientists made pronouncements, their refusal to admit mistakes, and to give any credence to the sheepfarmers' knowledge. Lastly, the scientists and bureaucrats displayed no sensitivity to the need for flexible decision-making by sheepfarmers.

The public face of certainty and the refusal to acknowledge mistakes contrasted with what farmers saw of the scientists at first hand as they descended on their farms in droves to carry out monitoring, sampling, field analysis and the like. The farmers watched the scientists debate where and how to take samples; they witnessed the variabilities of the readings, often over small distances; they noticed how hard it was to obtain a consistent calibration of background levels of radiation and the large number of variables that needed to be controlled. What the farmers were witnessing here was the normal open-ended nature of golem science which contrasted with the scientists' presentation of self.

One farmer described to Brian Wynne how his doubts set in as he watched his sheep being monitored for radioactivity:

> Last year we did five hundred [sheep] in one day. We started at 10.30 and finished at about six. Another day we monitored quite a lot and about 13 or 14 of them failed. And he [the person doing the monitoring] said, 'now we'll do them again' – and we got them down to three! It makes you wonder a bit . . . it made a difference . . . when you do a job like that you've got to hold it [the counter] on its backside, and sheep do jump about a bit.
>
> (Wynne, 1996, p. 33)

A good example of the scientists overlooking the expertise of the farmers was evident during one episode of experimentation on a farm. Scientists were looking for ways of getting rid of the radioactive caesium by absorbing it in other minerals. Differing concentrations of one possible mineral, bentonite, were spread on the ground in fenced-in plots. Sheep which had grazed on the plots were then tested for contamination and the results compared with a control group of sheep allowed to graze on the fells. The farmers criticized the experiment by pointing out that usually sheep grazed over *open* tracts of fell land and that if they were fenced in they would 'waste' (go out of condition). Such criticisms were ignored. Later, however, the farmers were vindicated when the experiments were abandoned for exactly the reasons the farmers had given.

Similarly, in the early days of the crisis, the scientists overlooked farmers' local knowledge of the lie of the land and their observations of where water accumulated and thus where radiation hot-spots were likely to form. The scientists seemed blind to the farmers' own expertise of the natural world where they lived and worked.

The biggest source of antagonism between the two groups resulted from the lack of understanding which officials displayed about fell sheepfarming as a way of life. The initial advice given to the farmers was to delay selling their lambs until the radiation had declined, but this showed a lack of familiarity with the subtleties of producing sheep for sale at a market. Choosing the right moment to take a lamb to market lies at the core of the farmers' skills, as Brian Wynne notes:

> The critical elements of the decision are flexibility and adaptability to the unexpected. Following variable winter conditions, several hundred lambs born over a period of six weeks from mid-April until the end of May will fatten to market readiness at different times from August to October. . . . Timing of the sales is critical because lambs rapidly go out of condition and lose market value if allowed to become overfat or underfed. This decision process requires highly flexible judgments: whether to take any lambs at all; if lambs are taken, how many and which ones; and to which of several nearby markets to take them. Complex craft judgements of trends in prices, rates of finishing of other lambs, pasture conditions, disease buildup, condition of the breeding ewes for mating for next year's lamb crop, need for money, and many other dynamic factors partly or fully beyond the control of the farmers enter into these decisions. It is an informal but highly refined process of expert judgment. The nature of this process runs completely counter to the rigidity of the bureaucratic method for dealing with the crisis: Sell the lambs later.
>
> (Wynne, 1989, p. 33)

It was not that sheepfarmers resented intervention, what they resented was the officials' refusal to come to terms with the complexity of the area in which they were intervening and the officials' refusal to credit their practical expertise.

In response to the overgrazing problem, on 13 August 1986 the government finally permitted farmers to move contaminated sheep from restricted areas and sell them as long as the lambs were marked by blue paint. All sheep in the restricted area had to be tested for contamination. Slaughtering the marked sheep for market was strictly forbidden until the original area they came from was derestricted. The sheep could, however, be sold on at loss to lowland farmers elsewhere. Contamination levels in sheep could be lowered by grazing sheep in the valleys instead of on the highly contaminated fells.

For most farmers this was not an option as the valley grass was needed for winter hay and silage crops; once grazed on the grass grows back quite slowly.

The technical experts dealt with this problem in a way which again demonstrated their lack of familiarity with fell farming. They first advised the farmers to keep the sheep in the valleys. This is wildly unrealistic for the reasons just given. They then suggested that the sheep could be fed imported feed 'such as straw'. A typical reaction to this suggestion was as follows:

> [The experts] don't understand our way of life. They think you stand at the fell bottom and wave a handkerchief and all the sheep come running . . . I've never heard of a sheep that would even look at straw as fodder. When you hear things like that it makes your hair stand on end. You just wonder, what the hell are these blokes talking about?
>
> *(Wynne, 1989, p. 34)*

The restrictions required farmers to notify MAFF officials at least five days in advance of their intention to sell sheep and to identify the market where they would be sold. Then a monitoring session had be arranged in advance. The lambs sometimes needed to be gathered and returned to the fells two or three times before they passed a monitoring session, a source of further frustration. Brian Wynne takes up the story again:

> Although the farmers accepted the needs for restrictions, they could not accept the experts' apparent ignorance of the effects of their approach on the normally flexible and informal system of hill farm management. This experience of expert knowledge being out of touch with practical reality and thus of no validity was often repeated with diverse concrete illustrations in interviews. Many local practices and judgments important to hill farming were unknown to experts, who assumed that scientific knowledge could be applied without adjusting to local circumstances. *(p. 34)*

THE SELLAFIELD FACTOR

The radiation from Chernobyl was not spread uniformly. One remarkable feature of the maps of the spread in the UK is that the highest levels in Cumbria form a crescent around the Sellafield nuclear reprocessing plant. This plant has a controversial history with a series of allegations of malpractice, radiation leaks and unexplained clusters of leukaemia cases in the nearby population. It is also where, prior to Chernobyl, the world's worst nuclear accident occurred. In 1957 a reactor caught fire and burned for three days. Milk from an area as large as 200 square miles had to be thrown away for fear of contamination. The accident was shrouded in secrecy and crucial data on its effects may never have been collected.

Because of the proximity of Sellafield to the Chernobyl fallout, local farmers were naturally suspicious that the two were linked. In one scenario, contamination from Sellafield had lain on the fells unnoticed until the Chernobyl disaster drew attention to it. As one farmer told Wynne:

> There's another thing about this as well. We don't live far enough away from Sellafield. If there's anything about we are much more likely to get it from there! Most people think that around here. It all comes out in years to come; it never comes out at the time. Just look at these clusters of leukaemia all around these places. It's no coincidence. They talk about these things coming from Russia, but it's surely no coincidence that it's gathered around Sellafield. They must think everyone is completely stupid.
>
> *(Wynne, 1996, p. 30)*

These suspicions were enhanced by the secrecy surrounding the 1957 fire. As another farmer reported:

> It still doesn't give anyone any confidence, the fact that they haven't released all the documents from Sellafield in 1957. I talk to people every week – they say this hasn't come from Russia! People say to me every week, 'Still restricted eh – that didn't come from Russia lad! Not with that lot [Sellafield] on your doorstep'.
>
> *(p. 30)*

And yet another farmer drew upon the perceived misinformation surrounding the Sellafield fire:

> Quite a lot of farmers around believe it's from Sellafield and not from Chernobyl at all. In 1957 it was a Ministry of Defence establishment – they kept things under wraps – and it was maybe

much more serious than they gave out. Locals were drinking milk, which should probably never have been allowed – and memory lingers on.

(p. 31)

In some cases farmers were convinced they had direct evidence of a link between Sellafield and the contamination. One farmer who lived near Sellafield reported:

If you're up on the tops [of the fells] on a winter's day you see the top of the cooling towers, the steam rises up and hits the fells just below the tops. It might be sheer coincidence, but where the [radiation] hot spots are is just where that cloud of steam hits – anyone can see it if they look. You don't need to be a scientist or be very articulate . . .

(p. 31)

Wynne reports that he also talked to one farmer who remembered walking on deposited ash in the wake of the 1957 fire.

The way that the scientific experts reacted to such suspicions further damaged their credibility in the eyes of the farmers. The scientific response was to state unequivocally that the two events were unrelated. The Sellafield and Chernobyl discharges contained two isotopes of caesium, caesium 137 and caesium 134, with very different half-lives. The half-life of caesium 137 is about thirty years while that of caesium 134 is one year. This means that over time the ratio of the amount of caesium 137 to caesium 134 should increase. The ratios found in Sellafield and Chernobyl contamination should be very different. The deposits found on the fells matched the expected value of the ratio, thus revealing the so-called Chernobyl 'fingerprint'. The experts proclaimed that the contamination was unambiguously produced by Chernobyl. But this did not satisfy the farmers.

Again Brian Wynne takes up the story:

To the farmers, this distinction was highly theoretical. Interviews with the farmers revealed a widespread belief that contamination from Sellafield had existed unrecognized since well before 1986. Farmers believed for several reasons that Sellafield radiation contributed to their high levels of contamination. First, the isotope ratio distinction could not be demonstrated to them. They were simply asked to believe the same experts who had shown themselves to be equally confident but wrong about the rate of decline of the contamination. The false certainty of the scientists was frequently cited as a sign of their lack of credibility. Second, the farmers were well aware of the fantastic variability of contamination over very small distances on their own farms; yet, they saw these variations processed in public scientific statements into average figures with no variability or uncertainty. If these measurements could be so misrepresented, perhaps the isotope ratios could also be more variable than what the experts would publicly acknowledge.

(Wynne, 1989, p. 36)

The farmers' suspicions about the reliability of the isotope ratios turned out to be warranted. It became clear as scientists refined their understanding of the contamination that, indeed, only about half of the observed radioactive caesium came from Chernobyl, with the rest originating in 'other sources' such as weapons testing fallout and the 1957 Sellafield fire.

It was often the very same experts who poured scorn on any link with Sellafield, who had earlier asserted with similar confidence that there would be no effects from the Chernobyl cloud, and furthermore that the subsequent restrictions would be short lived. The farmers were unimpressed by what they saw as scientists' overconfidence and false certainty:

My theory, which is probably as good as anyone else's is this: we don't know . . . They keep rushing to conclusions before the conclusion has been reached – you understand what I'm saying? They'd have been far better to keep their traps shut and wait.

(Wynne, 1996, p. 32)

CONCLUSION

What did the farmers make of their rather unsatisfactory encounter with the experts? As Wynne notes, they concluded that the scientists must have fallen victim to political pressure from the government. According to the farmers, the authorities were trying to cover up. In one scenario it is alleged that the authorities were waiting for Chernobyl to occur in order to give them the perfect excuse to pass off the previous unacknowledged contamination from the Sellafield plant. Such conclusions gained credibility whenever government heavy-handedness was encountered, such as a famous incident during the height of the crisis when MAFF officials refused to allow an American TV team to film the lively debate at a public meeting of local farmers. As he left the meeting, the TV producer commented in a loud and angry voice that his team had received more open treatment in (pre-glasnost) Poland!

Deep suspicions of scientists stem from the flip-flop thinking about science which we cautioned against in *The Golem*. Flipping to and fro between science being all about certainty and science being a political conspiracy is an undesirable state of affairs. That science is about certainty was the view first encountered by the farmers as scientists scrambled to react to the Chernobyl crisis. Rather than admitting to uncertainties, they made over-confident claims which, in the long term, were unsustainable. When the retractions eventually came, this encouraged the farmers to flip to a new view of science which was that the scientists were simply at the beck and call of their government masters.

Both views of scientific expertise are wrong. If scientists had treated the farmers as a group with relevant expertise in some areas, then both groups could have learned to value the others' contribution and seen the limitations of their own claims. Such an attitude would lead to a healthier climate in which to resolve matters of public concern.

In Britain the official response to public health risks has traditionally been paternalistic reassurance. The government judges that the danger of panic usually outweighs any real risk to its citizens. Thus their job is taken to be to allay public fears. This response can be seen at work in many incidents of public concerns over health such as the risk of salmonella poisoning from eggs, and more recently in the 'mad cow' episode where, despite mounting evidence of the link between eating beef and a degenerative brain disease, government scientists continually played it down. One of the dangers is that science as well as public health will be damaged.

A precondition for a more stable relationship between experts and their different publics must be that a notion similar to golem science should be given more currency. Without it we face instability in our political institutions and mass disenchantment with the very advice that we so desperately need – that of experts.

7

ACTing UP: AIDS cures and lay expertise

On 24 April 1984, Margaret Heckler, US Secretary of Health and Human Services, announced with great gusto at a Washington press conference that the cause of AIDS had been found. A special sort of virus – a retrovirus – later labelled as HIV, was the culprit. Vaccinations would be available within two years. Modern medical science had triumphed.

Next summer, movie star Rock Hudson died of AIDS. The gay community had lived and died with the disease for the previous four years. Now that the cause of AIDS had been found and scientists were starting to talk about cures, the afflicted became increasingly anxious as to when such cures would become available. Added urgency arose from the very course of the disease. The HIV blood test meant lots of seemingly healthy people were facing an uncertain future. Was it more beneficial to start long-term therapy immediately or wait until symptoms appeared? Given the rapid advance in medical knowledge about AIDS and the remaining uncertainties (even the cause of AIDS was a matter of scientific debate), was it better to act now with crude therapies or wait for the more refined treatments promised later?

AIDS: THE 'GAY PLAGUE'

AIDS is not confined to homosexuals, but in the US it was first described in the media as the 'gay plague' and gays as a community were quick to respond to its consequences. The gay community in the US is no ordinary group. The successful campaigns for gay rights in the sixties and seventies left them savvy, street-wise and well-organized. The large numbers of well-educated, white, middle-class, gay men added to the group's influence. Although Mainstreet America might still be homophobic, there were sizeable gay communities in several big cities, with their own institutions, elected officials and other trappings of political self-awareness.

Being gay had become relatively more legitimate, to some extent replacing earlier views of homosexuality which treated it as a disease or deviant condition. AIDS now threatened to turn the clock back and stigmatize gay men once more. In the public's mind, the disease was a biblical judgement brought on by gays' reckless promiscuity; with Reagan in power, and the right wing in ascendancy, AIDS became a way to voice many prejudices. For instance, conservative commentator, William F. Buckley Jr., proposed in a notorious *New York Times* op-ed piece in 1985:

> ... everyone detected with AIDS should be tattooed in the upper forearm to protect common-needle users, and on the buttocks, to prevent the victimization of other homosexuals...
> *(Epstein, 1996, p. 187)*

The charged atmosphere was evident in early community meetings held in the Castro district, at the heart of San Francisco's gay community. Captured movingly in the film *And the Band Played On* (based upon Randy Shilts's book of the same name), the gay community agonized over the difficult decision to close the bath houses – one of the most potent symbols of 1970s-style gay liberation. AIDS struck deep at the core institutions and values of the newly emancipated.

Grass-root activist organizations soon arose, dedicated to acquiring information about AIDS and how to fight it. Those who tested 'positive' for HIV were told to expect several years of normal life before the onset of the disease. AIDS activism not only fitted the psychological, physical and political circumstances in which people found themselves, but also, unlike many forms of activism, it promised direct benefits in terms of better medications and, perhaps, a cure.

The gay community was predisposed to be sceptical towards the world of science and medicine, especially as homosexuality had for years been labelled as a medical condition. To intervene in the AIDS arena the community would have to deal with some very powerful scientific and medical institutions. As we shall see, AIDS activists turned out to be remarkably effective in gaining and spreading information about the disease and its treatment. They also contributed to the scientific and medical debate to such an extent that they came to play a role in setting the AIDS research agenda and on occasion doing their own research. How such expertise was acquired by a lay group and applied so effectively is a remarkable story.

We tell this story in two parts. In Part I we trace some of the science of AIDS and document historically how the activists increasingly came to play a role in AIDS research. Part I concludes with the first official approval for use of an AIDS drug where the research was conducted to a significant extent by 'lay experts'. In Part II we further document the successes of the activists focussing in on one particularly influential group – ACT UP. We show how the lay activists acquired and refined their expertise such that they were able to mount a thorough-going political and scientific critique of the way that AIDS clinical trials were conducted. This critique by and large became accepted by the medical establishment.

PART I

A VACCINE IN TWO YEARS?

From the earliest moments of the AIDS epidemic it was apparent that misinformation was widespread. There were moral panics amongst the public about how AIDS could be transmitted. More significantly, early public pronouncements about the prospects for a cure were exaggerated. Scientists who had sat in on Heckler's press conference, winced when she talked about a vaccine being available within two years. At that point, reasonably effective vaccines had been developed for only a dozen viral illnesses and the most recent, against hepatitis B, had taken almost a decade to bring to market. Dr Anthony Fauci, head of the National Institute of Allergy and Infectious Diseases (NIAID), was more circumspect when he told the *New York Times*, a few days after Heckler's announcement, that:

> To be perfectly honest . . . we don't have any idea how long it's
> going to take to develop a vaccine, if indeed we will be able to
> develop a vaccine.
>
> *(Epstein, 1996, p. 182)*

Viruses invade the genetic code of the cell's nucleus – DNA – trans-
forming each infected cell into sites for the production of more virus.
Viruses, in effect, become part of each of the body's own cells. This is
quite unlike bacteria – cell-sized foreign bodies more easily identifiable and treatable by drugs such as antibiotics. To eliminate a virus
requires that every infected cell be destroyed without, at the same
time, harming the healthy cells. Worse, in the continual process of
virus reproduction it is likely that genetic mutations will arise; this
makes a viral disease even harder to treat.

THE PROMISE OF ANTI-VIRAL DRUGS

HIV differs from normal viruses in that it is a *retrovirus*. It is
composed of RNA (ribonucleic acid) rather than DNA
(deoxyribonucleic acid). Normally viruses work by turning cells into
virus factories using the blueprint of the original viral DNA. The
virus's DNA is copied into RNA which is then used to assemble the
proteins forming the new viruses. When retroviruses were first discovered they presented a problem. If they were made solely of RNA
how did they replicate? The answer was found in an enzyme known
as 'reverse transcriptase' which ensured that the RNA could be
copied into DNA. The presence of this enzyme offered the first
opportunities for a cure. If you could find an anti-viral agent that
eliminated reverse transcriptase, then perhaps HIV could be stopped
in its tracks. Some anti-viral drugs showed early promise in killing
HIV *in vitro* (literally, 'in glass', that is, outside of the body, in a test
tube).

Infection with the HIV virus can be diagnosed by a blood test – the
infected person being known as 'HIV positive'. Symptoms of the
disease AIDS may not appear until many years later. 'Full blown
AIDS' consists of a variety of opportunistic infective diseases which
develop because the body's immune system can no longer combat
them. 'Helper T cells', which are crucial in fighting such opportunis-

tic diseases, become depleted. Even if HIV could be killed *in vivo* ('in life', inside the body) a cure for AIDS might not necessarily follow. Perhaps long-term damage to T cells would have already occurred at an earlier stage of infection. Perhaps HIV infection interfered with autoimmune responses by some unknown route which meant that the immune system as a whole lost its ability to tell the difference between body cells and invaders.

In any event the path to a cure was likely to be long. An anti-viral compound had to be found which could be administered to humans in safe and clinically effective doses without producing damaging side effects. Its effectiveness had to be confirmed in controlled clinical trials with large numbers of patients. Lastly, it had to meet legal approval before being offered for widespread use.

CLINICAL CONTROLLED TRIALS AND THE FDA

Since the thalidomide scandal (thalidomide was developed as a sedative which was also used to treat morning sickness in pregnant women and later was found unexpectedly to cause severe birth defects) the Food and Drug Administration (FDA) required that new drugs be extensively tested before approval. Three phases of randomized control testing were demanded. In Phase I a small trial was necessary to determine toxicity and an effective dosage. In Phase II a larger, longer trial was carried out to determine efficacy. Phase III required an even larger trial to compare efficacy with other treatments. This process was costly and time consuming – typically it could take a new drug six to eight years to clear all the hurdles.

In October 1984, the first International Conference on AIDS took place in Atlanta, Georgia. These conferences, attended not only by scientists and doctors but also by gay activists and media personalities, became annual milestones. It was reported that small-scale trials had begun with six promising anti-viral drugs, including one called 'ribavirin'. But, Phase I trials were still a long way off. 'We have a long way to go before AIDS is preventable or treatable,' Dr Martin Hirsch of Massachusetts General Hospital concluded in reviewing the conference, 'but the first steps have been taken, and we are on our way.' (Epstein. 1996, p. 186)

BUYERS CLUBS

People dying of AIDS, and their supporters, were impatient with this kind of caution. They were desperate for anything, however unproven, to try and halt the march of the deadly disease, and they soon began to take matters into their own hands. Ribavirin was reported to be available at two dollars a box in Mexico. Soon this and other anti-viral drugs were being smuggled over the border into the US for widespread resale to AIDS sufferers. Illicit 'buyers clubs' started to flourish. Wealthy gay patients became 'AIDS exiles', moving to Paris, where another anti-viral agent, not approved in the US, was available.

Embarrassed by media stories about 'AIDS exiles', such as Rock Hudson, the FDA announced that new anti-viral drugs under test would be made available under a long-standing rule for 'compassionate use'. This meant that doctors could request experimental drugs for their terminally ill patients as a matter of last recourse.

PROJECT INFORM

The San Francisco gay community was a focal point for activism. Project Inform, a leading activist research group, was founded by Bay area business consultant, former seminary student, and ribavirin smuggler, Martin Delaney. The aim was to assess the benefits to be gained from new experimental drugs:

> 'No matter what the medical authorities say, people are using these drugs,' Delaney told reporters skeptical of the idea of community-based research. 'What we want to do is provide a safe, monitored environment to learn what effects they are having.'
>
> *(Epstein, 1996, p. 189)*

Although Delaney had no scientific background, he had personal familiarity with one of the key issues in the up-coming debate: who assumes the risk a patient faces in taking experimental drugs, the patient or the doctor?

Delaney had previously participated in an experimental trial of a new drug to treat hepatitis. The drug had worked for him but side effects had led to damage to the nerves in his feet. The trial was terminated and the treatment never approved because it was thought the drug was too toxic. Delaney, however, considered it a 'fair bargain' (Epstein. 1996, p. 189) as his hepatitis had been cured.

The prevailing trend in US clinical trials was to protect patients from harm. In 1974, Congress had created the National Commission for the Protection of Human Subjects with strict guidelines for research practices. This was in response to a number of scandals where patients had been unknowingly subject to experimentation. The most notorious was the Tuskegee syphilis study where, for years, poor black sharecroppers had been denied treatment so researchers could monitor the 'natural' course of the disease. Delaney, in pushing for patients to be given the right to do themselves potential harm with experimental treatments, seemed to be taking a backward step.

THE TRIALS OF AZT

The efforts of the activists to get more patients into experimental drug treatment programmes reached a peak in 1985 when, finally, it seemed that a promising anti-viral agent had been found. AZT (azidothymidine) had been developed originally to fight cancer. In this use the drug had failed. It had sat for years on the shelf at Burroughs Wellcome, the North Carolina-based subsidiary of the British pharmaceutical company, Wellcome. In late 1984, the National Cancer Institute (NCI) had approached leading pharmaceutical companies to send them any drugs that had the potential to inhibit a retrovirus and AZT was dusted off. In February 1985 AZT was found to be a reverse transcriptase inhibitor with strong anti-viral activity. A Phase I trial was immediately carried out. The six-week study of nineteen patients showed AZT kept the virus from replicating in fifteen of the patients, boosted their T cell counts, and helped relieve some of their symptoms. In transcribing the virus's RNA into DNA, AZT appeared to fool the reverse transcriptase into using it, in place of the nucleoside it imitated. Once AZT was added to the growing DNA chain, reverse transcriptase simply ceased to work and the virus stopped replicating. The problem with AZT was that, since it stopped DNA synthesis in the virus, there was every reason to believe that it might have harmful effects on DNA in healthy cells.

The NCI researchers were cautious in reporting their results because of the 'placebo effect'. It is well known that patients given dummy pills (placebos), which they are told are genuine, often feel better. In some mysterious way the psychological effect of taking a pill which patients believe will do them good, even though the pill has no medicinal qualities, actually produces a feeling of well-being and on some occasions even an improvement in health. Perhaps the effects reported by the NCI researchers were artifacts produced by patients' knowledge and expectation that AZT would help cure them? Although noting the immune and clinical responses to the drug, they warned about the possibility of a strong placebo effect. NCI called for a long-term double-blind controlled placebo study to be carried out in order to better assess the potential of AZT.

Funded by Burroughs Wellcome, plans went ahead to conduct this new trial at a number of different locations. At this time the testing of new AIDS drugs became more complicated because, with the help of $100 million in funding, NIAID started to set up its own network of centres to evaluate and test a variety of putative new AIDS drugs, including AZT. All this was done under the leadership of NIAID's head, Anthony Fauci. Creating the new centres took some time as a whole series of new research proposals and principal investigators had to be vetted. Time was something most AIDS patients did not have.

AIDS activist John James founded a San Francisco newsletter called *AIDS Treatment News*. *AIDS Treatment News* went on to become the most important AIDS activist publication in the United States. James was a former computer programmer with no formal medical or scientific training.

In the third issue of *AIDS Treatment News* James reported that large-scale studies of AZT were still months away and that even if all went well it would take another two years before doctors would be able to prescribe AZT. He estimated that at a death rate of ten thousand a year, which was expected to double every year, a two-year delay would mean that three quarters of the deaths which ever would occur from the epidemic would have occurred, yet would have been preventable.

In James's view a new task faced gay activists and AIDS organizations:

> So far, community-based AIDS organizations have been uninvolved in treatment issues, and have seldom followed what is going on . . . With independent information and analysis, we can bring specific pressure to bear to get experimental treatments handled properly. So far, there has been little pressure because *we have relied on experts* to interpret for us what is going on. They tell us what will not rock the boat. The companies who want their profits, the bureaucrats who want their turf, and the doctors who want to avoid making waves have all been at the table. The persons with AIDS who want their lives must be there too. [emphasis added]
>
> *(Epstein, 1996, p. 195)*

James did not regard AIDS researchers as being incompetent or evil. It was rather that they were too bound up in their own specialities and too dependent on bureaucratized sources of funding to be able to present and communicate a full and objective picture of what was going on. James believed that lay activists could become experts themselves:

> Non-scientists can fairly easily grasp treatment-research issues; these don't require an extensive background in biology or medicine.
>
> *(p. 196)*

As we shall see, James's optimism was not entirely misplaced.

Meanwhile the Phase II testing of AZT started. On 20 September 1986, a major study made headline news when it was ended early. AZT proved so effective that it was considered unethical to deny the drug to the placebo-taking control group. Dr Robert Windom, the assistant secretary for health, told reporters that AZT, 'holds great promise for prolonging life for certain patients with AIDS' (Epstein, 1996, p. 198) and he urged the FDA to consider licensing AZT as expeditiously as possible. With FDA and NIH support, Burroughs Wellcome announced that it would supply AZT free of charge to AIDS patients who had suffered the most deadly infectious disease, a particularly virulent form of pneumonia (PCP), during the past 120 days. Under pressure from AIDS patients and doctors who considered the criterion too arbitrary, this programme was expanded to include any of the 7,000 patients who had suffered PCP at any point.

On 20 March 1987, without a Phase III study, and only two years after first being tested, the FDA approved AZT for use. The cost of AZT treatment for a single patient was eight to ten thousand dollars a year (meaning it could only be used in wealthy western countries), and there is little doubt that Burroughs Wellcome made millions of dollars profit from the drug.

EQUIPOISE

In stopping the Phase II trial of AZT early the researchers had faced a dilemma. It made AZT more quickly available for everyone, but the chance of assessing its long-term effects under controlled conditions had been lost. The state of uncertainty as to which of the two arms in a clinical controlled study is receiving the better treatment is referred to as 'equipoise'. It is unethical to continue a trial if one treatment is clearly superior. In the case of AZT, the Phase II study had been 'unblinded' early by NIH's Data and Safety Monitoring Board. They concluded that equipoise no longer held: statistical evidence showed a treatment difference between the two arms of the study.

Equipoise sounds fine as an ideal, but it is not clear how easy it is to realize in practice. Are researchers ever in a genuine state of uncertainty? Even at the start of a controlled trial there must be some evidence of a drug's effectiveness or it would not have been tested in the first place. This dilemma famously faced Jonas Salk. He was so certain that his new polio vaccine worked that he opposed conducting a double-blind placebo study. Such a study, in his view, would mean that some people would unnecessarily contract polio. Salk's position was challenged by other researchers who claimed that, in the absence of such a study, the vaccine would not achieve broad credibility amongst doctors and scientists (Epstein, 1996, p. 201). It is obvious that the idea of equipoise involves complex social and political judgements concerning the credibility of a drug; furthermore such judgments are made by *researchers*, on behalf of *patients*.

PATIENTS AS BODY COUNTS

Patients in clinical trials are not passive research subjects. In the US, clinical trials have always been used by patients to get early access to experimental drugs. With AIDS activists making available information about new drugs almost as soon as they left the laboratory bench, patients clamoured to take part in AIDS clinical trials.

Two facets of the AZT trials were particularly disturbing to AIDS activists. Because the control group received placebos, this meant that in the long run the only way to tell whether a trial was successful or not was whether the body count in the placebo arm of the study was higher than in the other arm. In blunt terms, a successful study required a sufficient number of patients to die. This they considered unethical. A second criticism was of the rigid protocols of such studies; these forbade participants to take any other medication, even drugs which might prevent deadly opportunistic infections setting in.

Researchers engaged in such trials were quick to point out that the use of placebos was often the fastest route to learning about the effectiveness of a new drug and thus, in the long run, saved lives. They cited cases where claims had been made for the benefits of new drugs which randomized control trials had later shown to be useless and sometimes harmful. In response, activists pointed out that there were other options for controlled trials which did not use placebos. For instance, data from treatment groups could be compared with data from matched cohorts of other AIDS patients, or patients in studies could be compared against their own medical records. Such procedures were increasingly used in cancer research.

With the growing AIDS underground supplying drugs to patients in desperate circumstances, how accurate was the ideal scenario of the perfectly controlled clinical trial anyway? Although researchers did independent tests to assure 'compliance' (e.g. monitoring patients' blood for illicit medications), and claimed that, in general, patients did follow the protocols, the word from the real world of AIDS trials was somewhat different. Here is how Epstein describes matters:

> . . . rumors began to trickle in from various quarters: some patients were seeking to lessen their risk of getting the placebo by pooling

their pills with other research subjects. In Miami, patients had
learned to open up the capsules and taste the contents to distinguish
the bitter-tasting AZT from the sweet-tasting placebo. Dr. David
Barry, the director of research at Burroughs Wellcome, complaining
implausibly that never before in the company's history had any
research subject ever opened up a capsule in a placebo-controlled
trial, quickly instructed his chemists to make the placebo as bitter as
AZT. But patients in both Miami and San Francisco were then
reported to be bringing their pills in to local chemists for analysis.

(Epstein, 1996, p. 204)

REDEFINING THE DOCTOR – PATIENT RELATIONSHIP

'Non-compliance' is a long-standing concern amongst medical professionals. But what was happening in the case of AIDS patients was
something more radical – the patients, or, as they preferred to be
called, 'people with AIDS' – were renegotiating the doctor–patient
relationship into a more equal partnership. The feminist health,
self-help movements of the sixties and seventies had already shown
what could be achieved. With many gay doctors (some HIV positive)
in the gay communities, such a redefinition could be seen as being in
the interest of both doctors and patients.

The new partnership meant that patients had to start to learn the
language of biomedicine. Many were already well-educated (although not in the sciences), and no doubt this helped. Here is how
one AIDS patient described this process:

> . . . I took an increasingly hands-on role, pestering all the doctors:
> No explanation was too technical for me to follow, even if it took a
> string of phone calls to every connection I had. In school I'd never
> scored higher than a C in any science, falling headlong into
> literature, but now that I was locked in the lab I became as obsessed
> with A's as a premed student. Day by day the hard knowledge and
> raw data evolved into a language of discourse.
>
> *(Epstein, 1996 p. 207)*

Here is how doctors witnessed the same process:

> You'd tell some young guy that you were going to put a drip in his
> chest and he'd answer: 'No Doc, I don't want a perfusion inserted in

my subclavian artery, which is the correct term for what you proposed doing'.

(p. 207)

As AIDS patients acquired more and more information about the disease it became increasingly hard in clinical trials to separate out their role as 'patient' or 'research subject' from that of co-researcher.

The criticism by activists of placebo studies returned in 1987 when it became clear that AZT and other anti-viral agents might be more effective when given early, well before any symptoms appeared. A number of clinical trials began on early administration of AZT using placebo groups as controls. Researchers carrying out the trials felt the criticism of placebos in this case was balanced by the potentially toxic effect of AZT. It was uncertain whether AZT would be effective when given so early and *not* getting the toxic AZT in the placebo arm of the study might actually be beneficial to the subjects' health. Participants with AIDS, however, saw it differently, especially as they could not take their normal medication during the trial. As one person who discovered he was in the placebo arm of a trial said:

> Fuck them. I didn't agree to donate my body to science, if that is what they are doing, just sitting back doing nothing with me waiting until I get PCP or something.
>
> *(p. 214)*

This person also freely admitted that during the study he had taken illicit drugs smuggled in by the AIDS underground. Community doctors were appalled that seriously ill patients were not allowed to take their medicines because of the rigid protocols of such trials. Such doctors became increasingly sympathetic to activists as they tried to find ways out of the dilemma of being both patients and research subjects.

COMMUNITY-BASED TRIALS

Patient groups and community doctors eventually came up with a solution that was as simple as it was radical. They started working together to design their own trials. These would avoid the bureaucratic delays which faced the official trials; avoid what they considered the ethically dubious practice of using placebos; and, because of the close working relationship between doctors and patients, had the potential to assure much better compliance. In the mid-1980s two community-based organizations, one in San Francisco and one in New York, despite much official scepticism, started trials of new drugs. Such initiatives were particularly suitable for small-scale studies which did not require 'high tech' medical equipment. The new initiatives also found an unexpected ally in drug companies, who were getting increasingly impatient with the bureaucratic delays in the official NIAID tests.

One of the first successes of community-based trials was the testing of aerosolized pentamidine for the treatment of PCP. NIAID planned to test this drug but preparatory work had taken over a year. Meanwhile activists had pleaded with Fauci, head of NIAID, to draft federal guidelines approving the drug. Fauci refused, citing lack of data on its effectiveness. The activists came back from the meeting declaring 'We're going to have to test it ourselves'. And test it they did. Denied funding by NIAID, community-based groups in San Francisco and New York tested the drug without using any placebos. In 1989, after carefully examining the data produced by the community-based groups, the FDA approved the use of aerosolized pentamide. This was the first time in its history that the agency had approved a drug based solely on data from community-based research (Epstein, 1996, p. 218).

There is no doubt that the activists, in an unholy alliance with drug companies and deregulators, had been putting increasing pressure on the FDA. But none of this should take anything away from the activists' scientific achievements, which were very significant in terms of how we think about scientific expertise. As a lay group they had not only acquired sufficient expertise to become knowledgeable about the science of AIDS but they had, with the help of doctors, actually been able to intervene and conduct their own research. Furthermore, their research had been deemed credible by one of the most powerful scientific-legal institutions in the US – the FDA.

Having traced the early history of activist involvement and documented their first major success we now describe in Part II how the activists' own expertise was increasingly enlisted in the science of clinical trials. The radical critique of clinical trials offered by the activists was, as we shall see, eventually to become the established view, signalling a remarkable change in medical opinion as to how such trials should be conducted.

PART II

A C T U P

In the mid-1980s a new organization of AIDS activists took centre stage. ACT UP (the AIDS Coalition to Unleash Power) was soon to have chapters in New York and San Francisco as well as other big US cities. By the 1990s there were groups in cities in Europe, Canada and Australia as ACT UP became the single most influential AIDS activist group.

ACT UP practised radical street politics: 'No business as usual'. A typical ACT UP demonstration took place in fall 1988 on the first day of classes at Harvard Medical School. Equipped with hospital gowns, blindfolds, and chains, and spraying fake blood on the sidewalk to the chant of, 'We're here to show defiance for what Harvard calls "good science"!', the activists presented Harvard students with an outline of a mock AIDS 101 class. Topics included:

> *PWA's [People With AIDS] – Human Beings or Laboratory Rats?
> *AZT – Why does it consume 90 percent of all research when it's highly toxic and is not a cure?
> *Harvard-run clinical trials – Are subjects genuine volunteers, or are they coerced?
> *Medical Elitism – Is the pursuit of elegant science leading to the destruction of our community?
> *(Epstein, 1996, p. 1)*

One of ACT UP's political messages was that AIDS was a form of genocide by neglect. Because of the Reagan government's indifference and intransigence, AIDS was spreading with no cure in sight other than the highly toxic AZT. One of the first targets of ACT UP was the FDA – the 'Federal Death Agency', as it was branded by the activists. The culmination of a campaign of protest came in October

1988 when a thousand demonstrators converged on FDA head-quarters. Two hundred demonstrators were arrested by police wearing rubber gloves. The subsequent media attention and negotiations with the FDA meant that the government recognized for the first time the seriousness and legitimacy of the activists' case.

Unlike other activist protest movements, such as those in favour of animal rights, ACT UP did not see the scientific establishment as the enemy. In public they applied pressure via well-publicized protests; in private, however, they were quite prepared to engage with the scientists and argue their case. It was the scientists, indeed, to whom the activists increasingly turned their attention. The FDA had been a potent symbol in terms of mobilizing popular opinion, but what really mattered now was getting access to NIAID and the NCI and persuading them to conduct clinical trials in a different manner. It meant engaging with what the FDA called 'good science'.

TALKING GOOD SCIENCE

AIDS Treatment News heralded the new agenda by declaring, in 1988, 'the more important question is what treatments do in fact work, and how can the evidence be collected, evaluated, and applied quickly and effectively'. (Epstein, 1996, p. 227)

Over the next three years the activists developed a three-pronged strategy: (1) force the FDA to speed up approval of new drugs; (2) expand access to new drugs outside of clinical trials; and (3) transform clinical trials to make them 'more humane, relevant and more capable of generating trustworthy conclusions'. Points (2) and (3), in particular, represented a departure from the standard way of thinking about clinical trials. The normal way of enticing patients to take part in clinical trials was to create conditions under which bribery would be effective: access to drugs was restricted outside of the trials. In contrast, AIDS activists saw restricted access as the cause of many of the difficulties encountered in clinical trials. As Delaney argued:

> The policy of restriction . . . is itself destroying our ability to conduct clinical research . . . AIDS study centers throughout the nation tell of widescale concurrent use of other treatments; frequent cheating,

even bribery to gain entry to studies; mixing of drugs by patients to share and dilute the risk of being on placebo; and rapid dropping out of patients who learn that they are on placebo . . . Such practices are a direct result of forcing patients to use clinical trials as the only option for treatment . . . If patients had other means of obtaining treatment, force-fitting them into clinical studies would be unnecessary. Volunteers that remained would be more likely to act as pure research subjects, entering studies not only out of a desperate effort to save their lives.

(p. 228)

This was a clever argument because rather than rejecting the idea of clinical trials it suggested a way in which such trials could be made more reliable. To press this argument required the activists more and more to enter the specialist territory of the scientists and medics – in effect the activists had to tell the medical establishment how to run their trials better.

The activists, as has been pointed out, often started with no scientific background. Remarkably, they quickly acquired a new kind of reputation. They were seen by the physicians and scientists to possess formidable knowledge and expertise about AIDS and its treatment. Practising physicians were some of the first to encounter the new-found experts. Soon the activists found that physicians were turning to *them* for advice. The director of the New York City Buyers Club was quoted as saying:

When we first started out, there were maybe three physicians in the metropolitan New York area who would even give us a simple nod of the head . . . Now, every day, the phone rings ten times, and there's a physician at the other end wanting advice. [from] me! I'm trained as an opera singer.

(p. 229)

Some activists, of course, did have a medical, scientific or pharmacological background and such people soon became indispensable as teachers of raw recruits. But most of the leading figures were complete science novices. Mark Harrington, leader of ACT UP New York's Treatment and Data Committee, like many activists, had a background in the humanities. Before entering ACT UP he was a script writer:

The only science background that might have proved relevant was [what I had] when I was growing up: my dad had always subscribed to *Scientific American*, and I had read it, so I didn't feel that sense of intimidation from science that I think a lot of people feel.

(p. 230)

Harrington stayed up one night and made a list of all the technical words he needed to understand. This later became a fifty-three page glossary distributed to all ACT UP members.

Other activists were overwhelmed when they first encountered the technical language of medical science, but they often reported that, like learning any new culture or language, if they kept at it long enough things started to seem familiar. Here is how Brenda Lein, a San Francisco activist, described the first time she went to a local meeting of ACT UP:

And so I walked in the door and it was completely overwhelming, I mean acronyms flying, I didn't know *what* they were talking about . . . Hank [Wilson] came in and he handed me a stack about a foot high [about granulocyte macrophage colony-stimulating factor] and said, "Here, read this." And I looked at it and I brought it home and I kept going through it in my room and . . . I have to say I didn't understand a word. But after reading it about ten times . . . Oh this is like a subculture thing; you know, it's either surfing or it's medicine and you just have to understand the lingo, but it's not that complicated if you sit through it. So once I started understanding the language, it all became far less intimidating.

(p. 231)

The activists used a wide variety of methods to get enculturated in the science. These included attending scientific conferences, reading research protocols and learning from sympathetic professionals both inside and outside the movement. The strategy used was often that of learning, as one activist called it, 'ass-backwards'. They would start with one specific research proposal and they would then work back from that to learn about the drug mechanism and any basic science they would need. The activists considered it a *sine qua non* of their effective participation that they would need to speak the language of the journal article and the conference hall. In other words they saw the need to challenge the established experts *at their own game*. In this, it seems, they were remarkably effective – once researchers got used to their rather unusual appearance! As Epstein comments on activist Brenda Lein's involvement:

> Lein again: "I mean, I walk in with . . . seven earrings in one ear and a Mohawk and my ratty old jacket on, and people are like, "Oh great, one of those street activists who don't know anything . . ." But once she opened her mouth and demonstrated that she could contribute to the conversation intelligently, Lein found that researchers were often inclined, however reluctantly, to address her concerns with some seriousness.
>
> *(p. 232)*

Or, as one leading authority on clinical trials commented:

> About fifty of them showed up, and took out their watches and dangled them to show that time was ticking away for them . . . I'd swear that the ACT UP group from New York read everything I ever wrote . . . And quoted whatever served their purpose. It was quite an experience.
>
> *(p. 232)*

There is no doubt that some AIDS scientists in their initial encounters were hostile towards the activists. Robert Gallo the co-discover of HIV is reported as saying:

> I don't care if you call it ACT UP, ACT OUT or ACT DOWN, you definitely don't have a scientific understanding of things.
>
> *(p. 116)*

Gallo later referred to activist Martin Delaney as, 'one of the most impressive persons I've ever met in my life, bar none, in any field . . . I'm not the only one around here who's said we could use him in the labs.' Gallo described the level of scientific knowledge attained by certain treatment activists as 'unbelievably high': 'It's frightening sometimes how much they know and how smart some of them are . . .' (p. 338)

ACTIVISTS START TO WIN ALLIES

By 1989 the activists were beginning to convince some of the most powerful scientists of their case. No less a person than Anthony

Fauci, the head of NIAID, started a dialogue. Fauci told the *Washington Post*:

> In the beginning, those people had a blanket disgust with us . . . And
> it was mutual. Scientists said all trials should be restricted, rigid and
> slow. The gay groups said we were killing people with red tape.
> When the smoke cleared we realized that much of their criticism was
> absolutely valid.
>
> *(Epstein, 1996, p. 235)*

The weight of the activist arguments was finally realized at the Fifth International Conference on AIDS held in Montreal in June 1989. Protesters disrupted the opening ceremony, demonstrated against particular profit-hungry pharmaceutical companies, and presented formal posters on their views of drug regulation and clinical trials. Leading activists met with Fauci and enlisted his support for their notion of 'Parallel Track'. Under this scheme drugs would be made available to patients reluctant to enter a clinical trial while at the same time the trial would go ahead. Scientists worried whether this would mean less patients in trials, but patients did continue to enrol in trials even after Parallel Track was adopted.

Activists also sowed doubt about some of the formal rules governing randomized controlled trials. The crucial breakthrough again came at the Montreal conference where ACT UP New York prepared a special document critical of NIAID trials. Susan Ellenberg, the chief bio-statistician for the trials, recalled seeking out this document in Montreal:

> I walked down to the courtyard and there was this group of guys,
> and they were wearing muscle shirts, with earrings and funny hair. I
> was almost afraid. I was really hesitant even to approach them.
>
> *(p. 247)*

Ellenberg, on actually reading the document, found, much to her surprise, that she agreed with some of the activists' points. Back at her lab she immediately organized a meeting of medical statisticians to discuss the document further. This was apparently an unusual meeting. As she remarked:

> I've never been to such a meeting in my life.
>
> *(p. 247)*

Another participant said:

> I think anybody looking at that meeting through a window who could not hear what we were saying would not have believed that it was a group of statisticians discussing how trials ought to be done. There was enormous excitement and wide divergence of opinion.
>
> *(p. 247)*

So impressed were the statisticians by the activists' arguments that members from ACT UP and other community organizations were invited to attend regular meetings of this group. The debate was over whether 'pragmatic' clinical trials, which took into account the messy realities of clinical practice, might actually be more desirable scientifically. Here the activists tapped into a long-running controversy amongst bio-statisticians as to whether clinical trials should be 'fastidious' or 'pragmatic'. A pragmatic clinical trial works under the assumption that the trial should try and mirror real world untidiness and the heterogeneity of ordinary clinical practice patients. Such pragmatic considerations were already familiar to some bio-statisticians who had experience of cancer trials, where different and more flexible ways of thinking about trial methodology had already been instituted. The fastidious approach favoured 'clean' arrangements, using homogeneous groups. The problem with the fastidious approach was that although it might deliver a cleaner verdict, that verdict might not apply to the real world of medical practice where patients might be taking a combination of drugs.

THE EXPERTNESS OF LAY EXPERTISE

What particular expertise did the activists possess? Or did they just have political muscle? Scientists are predisposed to avoid political interference in their work, especially by untrained outsiders. Without any expertise to offer, the politicking of the activists would, if anything, have made the scientists more unsympathetic.

The activists were effective because they had some genuine expertise to offer and they made that expertise tell. In the first place, their long experience with the needs of people with AIDS meant that they were familiar with the reasons subjects entered studies and how they could best be persuaded to comply with protocols. Fauci described this as:

> an extraordinary instinct . . . about what would work in the community . . . probably a better feel for what a workable trial was than the investigators [had].
>
> *(Epstein, 1996, p. 249)*

Activists also had a particularly valuable role to play as intermediaries explaining to people with HIV and AIDS the 'pros and cons' of particular trials.

But the expertise went beyond this. By learning the language of science they were able to translate their experience into a potent criticism of the standard methodology of clinical trials. By framing their criticisms in a way that scientists could understand they forced them to respond. This was something which the sheepfarmers discussed in Chapter 6 were unable to do. The activists were lucky because at the very time they were raising these concerns some biostatisticians were reaching broadly similar conclusions themselves.

One of the most fascinating aspects of the encounter between the activists and the scientists was the give and take on both sides. For example, as the activists learned more and more about the details of clinical trials they started to see why, in some circumstances, a placebo study might be valuable. Thus, AIDS activist, Jim Eigo, in a panel discussion in 1991, acknowledged that although originally he had seen no need for placebos, he now recognized the virtues of using them in certain situations where a short trial could rapidly answer an important question.

While AIDS activists embraced the 'real world messiness' model of clinical controlled trials, some were chastened by the experience of carrying out real world research. Martin Delaney of Project Inform admitted, after conducting a controversial clinical trial without any placebos, 'The truth is, it does take a lot longer to come up with answers than I thought before' (Epstein, p. 258).

TEACHING OLD DOGS NEW TRICKS

The success of the activists' arguments about the need to make clinical trials more patient orientated was marked in October 1990 by the publication of two back-to-back 'Sounding Board' articles in the *New England Journal of Medicine*. One, by a group of prominent bio-statisticians, argued for restructuring the phases of the FDA approval process, dismissed the requirement of homogeneity in clinical trial populations, and called for more flexible entry criteria. It concluded by calling for patients to participate in the planning of clinical trials. The second article by a well-known Stanford AIDS researcher was titled, 'You *Can* Teach an Old Dog New Tricks: How AIDS Trials are Pioneering New Strategies' and took up similar themes of flexibility, and how to ensure each limb of a trial offered benefits to patients. Medical ethicists soon came on board supporting the new consensus about how AIDS trials should be conducted. And indeed, trials started to be conducted according to the protocols suggested originally by the activists. Another victory came when NIAID started to recruit an increasingly diverse population into trials.

By the next international AIDS conference the activists were so well accepted into the AIDS establishment that they spoke from the podium rather than shouting from the back of the room (Epstein p. 286). In a speech at the conference, Anthony Fauci announced:

> When it comes to clinical trials, some of them are better informed than many scientists can imagine.
>
> *(p. 286)*

The success of the activists in speaking the language of science had one paradoxical outcome, it meant that new generations of activists increasingly felt alienated from the older activists. Indeed splits and tensions appeared between 'expert lay' activists and 'lay lay' activists. As one New York activist reflected:

> ... there were people at all different points within the learning curve ... You'd have somebody ... who had AIDS, who knew a lot about AIDS, [but who], didn't know *anything* about AIDS research – you know nothing. And never had seen a clinical trial, didn't live in a city where they did clinical trials, on the one end – and then Mark Harrington and Martin Delaney on the other.
>
> *(p. 293)*

This division of expertise is exactly what we would expect given the golem model of science. Expertise is something that is hard won in practice. Someone with AIDS might be an expert on the disease as it affects patients, but this does not make them an expert on the conduct of clinical trials. It was the particular expertise of the activists which commanded the scientists' attention and if new activists were going to have any influence, they too would need to become experts.

Some AIDS activists, as they became increasingly enmeshed in the science of AIDS, even became 'more scientific than thou' as we might say, when it came to assessing treatments. On one famous occasion, well-known activists chastised a leading AIDS researcher for trying to make the best of clinical trial data by *post hoc* redrawing of the sample into sub-groups in order to claim some effect. Some activists refused to countenance alternative therapies for AIDS and some even enrolled in medical school in order to undertake formal scientific training.

The activists were not a homogeneous body, and splits and tensions arose between different groups and within groups. The New York activists in comparison to the San Francisco group were seen as being more closely aligned to orthodox science. But even the New York activists always maintained that a core part of their expertise lay in their being part of the community who were living with and dying from AIDS. It was their experiences of the world of the patients which gave them something which no medical expert could hope to acquire unless they themselves had AIDS or were part of the gay community.

Although activists went on through the 1990s to contribute to other issues such as debates over combination therapies and the role of surrogate markers in assessing the severity of AIDS, it is in the arena of clinical trials that their most stunning victories were achieved. In effect, a group of lay people had managed to reframe the scientific conduct of clinical research: they changed the way it was conceived and practised.

This success shows us that science is not something that only qualified scientists can do. Lay people can not only gain expertise but also, in some circumstances, they can have that expertise accorded the respect it deserves. Treating science and technology as forms of expertise, which are similar in principle to other forms of expertise, opens the door to understanding how this lay expertise is possible.

Just as lay people can gain expertise in plumbing, carpentry, the law and real estate, they can gain expertise in at least some areas of science and technology. And in some areas of science and technology, they may already have more relevant experience than the qualified experts. But the crucial issue as we have seen in this chapter is getting that expertise recognized as expertise. And it is this more than anything which the AIDS activists have been able to achieve.

Conclusion: the golem goes to work

The conclusion of *The Golem*, the first volume in this series, argued that the book had wide significance where science touched on matters of public concern. Here we deliver on that promise.

The chapter on the *Challenger* explosion shows the way that human error is taken to account for technological failure and shows how unfair it is to assign blame to individuals when the uncertainties are endemic to the system as a whole.

The *Challenger* enquiry is one case among many that reveals that when the public views the fruits of science from a distance the picture is not just simplified but significantly distorted. Nobel laureate Richard Feynman demonstrated on TV that when a piece of rubber O-ring was placed in a glass of iced water it lost resilience. This was at best trivial – the effect of low temperature on rubber was already well understood by the engineers. At worst it was a dangerously misleading charade – an acting-out of the most naive model of scientific analysis. The crucial question was not whether low temperature affected the O-rings but whether NASA had reason to believe this would cause them to fail. Feynman gives the impression that doubts can always be simply resolved by a scientist who is smart enough.

Feynman was encouraging us all to be enchanted by the power of the scientific method, but the enchantment can last only while we remain at a safe distance in time, space, and understanding, from technology in practice – distance lends enchantment in technology as well as science. The danger is always that enchantment is the precursor of disenchantment.

The chapter on the train and plane crashes, and the discussion of the Patriot missile reinforce the message that distance lends enchantment. Both show how biased the distanced view can be.

The chapter on the econometric modellers shows that the canonical model of science can mislead when it is applied uncritically. Social science has its own methods and not all of these are quantitative by any means. Where mathematisation is taken by social scientists to be the only key to sound conclusions, social science turns somersaults. In the case of the econometric modellers, we argue, the science – that is to say the expertise – is not found in the mathematical models, which are merely the litter from an elaborate training exercise, but in the judgements which emerge from long experience.

The most straightforward manifestation of the experimenter's regress – the 'technologist's regress' as we might rename it for the purpose of this volume – is found in the chapter about the origins of oil. Two wells were drilled, each designed to give an unambiguous answer to the question of where oil comes from, but both failed to do so. Of course, the technologist's regress is revealed in every chapter of the book.

In the conclusion of *The Golem* we discussed forensic science. Though we do not present a case study of forensic science here, it is worth noting that events in the world have overtaken our analysis. Remarkably, for almost a year, a worldwide television audience could watch the process of scientific evidence being taken apart during cross-examination in the O. J. Simpson trial. What were viewers and jurors to make of the conflicting claims made for DNA evidence? For the prosecution it provided the indisputable proof of Simpson's guilt; for the defence it was a tendentious filament of argument resting on undocumented laboratory practice and subject to the contingencies of police procedures. This was what *The Golem* foresaw.

The golem view of science shows not only why expert evidence is so easy to deconstruct in a courtroom but also how evidence can regain respect. That experts disagree is not a good reason to discount evidence, and it is now being realized that expert witnesses may not always be able to provide a complete description of their scientific procedures. In the Simpson trial, an expert witness for the prosecution, Dr Robin Cotton, argued that she could not be called to task for not being able to detail her practice in such a way as to show that it exactly followed a set of formal rules. Cotton argued that it was her expertise on which the court should rely, and Judge Ito accepted this. Within the model of expertise, this is not unreasonable.

Once more, the word of an expert, as an expert, must sway us only if the expertise applies to the problem in hand. Expertise can be used incorrectly. Risk analysis is one such example: each of us is entitled to our own preference about whether we would rather burn coal and risk death from lung disease or burn nuclear fuel and risk the death of whole communities from nuclear catastrophes. There are certain kinds of calculations that suggest that the number of deaths associated with breathing the products of burning fossil fuels is likely to be greater than the number of deaths associated with nuclear catastrophes. But the expertise used to make these predictions is expertise in well-specified calculations, it is not expertise in defining the acceptability of risks. Firstly, the dangers associated with nuclear power depend on the political climate – under conditions of widespread terrorism or anarchy, nuclear power becomes much more dangerous in a way that fossil fuel technology does not. And this means there are two dangers: the danger of nuclear explosion and the danger of the imposition of especially repressive political regimes. These factors cannot feature in the expert calculations. Secondly, the calculations cannot take into account preferences in respect of the distribution of risk. Perhaps we would prefer to see 100,000 deaths spread throughout a large population than 10,000 deaths concentrated in a small community downwind of a nuclear plant. The death of individuals is a normal part of life while the death of whole communities is not. Golem science and technology is a body of expertise, and expertise must be respected. But we should not give unconditional respect before we understand just what the expertise comprises and whether it is relevant. To give unconditional respect is to make science and technology a fetish.

In a case like public acceptability of risk we should draw the opposite conclusion to that we drew in the case of the economists. Both sets of experts were practised in doing calculations – in the case of the economists these exercises gave them expertise in the working of the economy; in the case of risk analysis the calculations do not provide expertise in risk acceptability.

As well as illustrating many of the claims made in the first volume, *The Golem at Large* has explored the boundaries of expertise. The chapters on the Chernobyl fallout and on AIDS cures have shown that these boundaries do not always coincide with the boundaries of formally certified training. Trained persons who try to do their science, technology, or medicine, by cleaving to strict formulaic procedures are often ineffective practitioners. On the other hand, uncertified and unformulated experience can be of enormous value. But, as we stress over and over again, this position does not justify the idea that one person's contribution is as good as another's. Expertise is the key, it is just that every now and again we find that it has been attained by an unorthodox route.

When Cumbrian farmers found themselves dealing with scientists in the aftermath of the Chernobyl disaster, it did nothing for the public reputation of science. The same is true of another recent incident that took place in Britain – the outbreak of BSE or 'mad cow disease'. Could the disease be passed between species – sheep to cows and vice versa? Could it be the cause of a new variant of a human brain disease? Was it safe to eat hamburgers? The contradictory answers coming from the scientific community, the attempts to conceal the whole truth, the political debacle as British beef was banned from export markets, the well-publicized collapse of British beef farming as indecisive policy followed indecisive policy: none of these were good for science. And yet in many ways the fault lay not with science, which was simply unable to deliver instant answers to some very subtle questions, but with the false expectations placed upon it. This allowed others to pass the blame to science whenever it suited them.

It is vital that the disillusion that follows such episodes does not become so widespread that science and technology are no longer valued; there is nowhere else to turn if our society is not to fall back into a dark age. We have to learn to have the right expectations. Neither science nor technology are the kind of 'higher superstitions', or quasi-religions, that many 'scientific fundamentalists' take them to be. The case studies in this book are meant to help to avoid the catastrophic flight from reason that could follow from an accumulation of incidents like the BSE debacle.

CONCLUSION

We offer no policies: we say neither 'eat hamburgers' nor 'don't eat hamburgers'; we do not press our readers to vote for those who would build anti-missile missiles or for those who would not; we do not demand that all drugs be tested in randomized control trials nor that all drugs trials rate ethical considerations above scientific testing. What we try to do is make space to think about these things. If the only available model of technological development sets perfection against failure with no middle road, then we will not be able to make sense of the technological world; perfection will never be attained and 'failure' will be the only conclusion left. This is the flip-flop thinking we warned against in *The Golem*.

In that first volume of the Golem series we likened frontier scientists to high school students testing the temperature of boiling water. Perhaps, as we say in the Introduction, a better metaphor for technology is the cook or gardener. No one who is clumsy in the kitchen or the garden shed can possibly doubt that there are those who are more expert than themselves. Yet no one expects the cook or gardener to produce a perfect soufflé or herbaceous border every time. That one gets better through practice is also clear; that there are a few virtuosos is also obvious; that not everyone's opinion on matters of cookery or gardening is as valuable as everyone else's is not a matter for dispute. Take away the mystery and the fundamentalism and we can see frontier technology as the application of expertise in trying circumstances. That is how we must find our way through the technological world.

References and further reading

Atkinson, R. (1994) *Crusade: The Untold Story of the Gulf War*, London: HarperCollins.

Bijker, W., Hughes, T. and Pinch, T. (1987) *The Social Construction of Technological Systems*, Cambridge, Mass.: MIT Press.

Burrell, A. and Hall, S. (1994) 'A Comparison of Macroeconomic Forecasts', *Economic Outlook*, Briefing Paper, 29–35, February 1994.

Cole, S. A. (1996) 'Which Came First, the Fossil or the Fuel?', *Social Studies of Science*, **26**, 733–66.

Cole, S. A. (1998) 'It's a Gas!', *Lingua Franca*, December 1997/January 1998, 11–13.

Collins H. M. (1985/1992) *Changing Order: Replication and Induction in Scientific Practice*, London and Beverley Hills: Sage. [Second Edition, Chicago: The University of Chicago Press.]

Collins, H. M. (1988) 'Public Experiments and Displays of Virtuosity: The Core-Set Revisited', *Social Studies of Science*, **18**, 725–48.

Collins, H. M. (1991) 'The Meaning of Replication and the Science of Economics', *History of Political Economy*, **23**, 1,123–43.

de la Billiere, P. (1994) *Looking for Trouble: SAS to Gulf Command, the Autobiography*, London: HarperCollins.

Epstein, S. (1995) 'The Construction of Lay Expertise: AIDS Activism and the Forging of Credibility in the Reform of Clinical Trials', *Science Technology & Human Values*, **20**, 408–37.

Epstein, S. (1996) *Impure Science: AIDS, Activism and the Politics of Knowledge*, Berkeley, Los Angeles and London: University of California Press.

Evans, R. J. (1997) 'Soothsaying or Science: Falsification, Uncertainty and Social Change in Macro-economic Modelling', *Social Studies of Science*, **27**, 3, 395–438.

Evans, R. J. (1997) 'What Happens Next? Can Economic Forecasters Foretell the Future?' Ph.D. Thesis. University of Bath.

Gieryn, T. F. and Figert, A. E. (1990) 'Ingredients for a Theory of Science in Society: O-Rings, Ice Water, C-Clamp, Richard Feynman and the New

York Times', in S. Cozzens and T. Gieryn (eds.) *Theories of Science in Society*, Bloomington: Indiana University Press.

Gooding, D. (1985) 'In Nature's School: Faraday as an Experimentalist', in D. Gooding and F. A. James (eds.) *Faraday Rediscovered: Essays on the Life and Work of Michael Faraday, 1791–1876*, London: Macmillan.

Kuhn, T. S. (1961) 'The Function of Measurement in Modern Physical Science', *Isis*, **52**, 162–76.

Lewis, M. (1989) *Liar's Poker*, London: Hodder and Stoughton.

MacKenzie, D. (1991) *Inventing Accuracy: A Historical Sociology of Ballistic Missile Guidance*, Cambridge, Mass.: MIT Press.

McCloskey, D. (1985) *The Rhetoric of Economics*, Madison, Wis.: University of Wisconsin Press.

McConnel, M. (1988) *Challenger: 'A Major Malfunction'*, London: Unwin.

Ormerod, P. (1994) *The Death of Economics*, London: Faber and Faber.

'Performance of the Patriot Missile in the Gulf War', Hearing before the Legislation and National Security Committee of the Committee on Government Operations, House of Representatives One Hundred and Second Congress, Second Session, April 7 1992.

Pinch, T. J. (1986) *Confronting Nature: The Sociology of Solar-Neutrino Detection*, Dordrecht: Reidel.

Pinch, T. J. (1991) 'How Do We Treat Technical Uncertainty in Systems Failure? The Case of the Space Shuttle *Challenger*', in T. La Porte (ed.) *Responding to Large Technical Systems: Control or Anticipation*, Dordrecht: Kluwer, 137–52.

Postol, T. A. (1991) 'Lessons on the Gulf War Experience with Patriot', *International Security*, **16**, 119–71.

Postol, T. A. (1992) 'Correspondence', *International Security*, **17**, 225–40.

Reports of the Seven Wise Men. HM Treasury.

Report of the Presidential Commission on the Space Shuttle Challenger Accident (1986). Five Volumes. Washington, June 6.

Shapin, S. (1988) 'The House of Experiment in Seventeenth-Century England', *Isis*, **79**, 373–404.

Shapin, S. (1994) *A Social History of Truth: Civility and Science in Seventeenth-Century England*, Chicago: The University of Chicago Press.

Stein, R. M. (1992) 'Correspondence: Patriot Experience in the Gulf War', *International Security*, **17**, 199–225.

Van Creveld, M. (1985) *Command in War*, Cambridge, Mass.: Harvard University Press.

Vaughan, D. (1996) *The Challenger Launch Decision: Risky Technology*,

Culture and Deviance at NASA, Chicago: The University of Chicago Press.

Wallis, K. F. (ed), Andrews, M. J., Fisher, P. G., Longbottom, J. A. and Whitely, J. D. (1987) *Models of the UK Economy: a Third Review by the ESRC MacroModelling Bureau*, Oxford: Oxford University Press.

Wynne, B. (1988) 'Unruly Technology: Practical Rules, Impractical Discourses and Public Understanding', *Social Studies of Science*, **18**, 147–67.

Wynne, B. (1989) 'Sheepfarming after Chernobyl: A Case Study in Communicating Scientific Information', *Environment*, **31**, 10–15, 33–9.

Wynne, B. (1996) 'Misunderstood Misunderstandings: Social Identities and Public Uptake of Science', in A. Irwin and B. Wynne, *Misunderstanding Science? The Public Reconstruction of Science and Technology*, Cambridge and New York: Cambridge University Press, 19–46.

Index